BRITISH WEIGHTS & MEASURES

*A History
from Antiquity
to the Seventeenth Century*

BRITISH WEIGHTS & MEASURES

*A History
from Antiquity
to the Seventeenth Century*

RONALD EDWARD ZUPKO

The University of Wisconsin Press

Published 1977
The University of Wisconsin Press
Box 1379, Madison, Wisconsin 53701

The University of Wisconsin Press, Ltd.
70 Great Russell Street, London

Copyright © 1977
The Regents of the University of Wisconsin System
All rights reserved

First printing

Printed in the United States of America

For LC CIP information see the colophon

ISBN 0-299-07340-8

To the memory of
Robert L. Reynolds

Contents

Figures	ix
Acknowledgments	xi
Introduction	xiii
1. ROMAN, ANGLO-SAXON, AND NORMAN LEGACIES	3
Roman Britain	3
Anglo-Saxon Britain	8
Norman Britain	14
2. MEDIEVAL TIMES: REFINEMENT AND PROLIFERATION	16
Angevin-Plantagenet Legislation	16
Lancastrian Legislation	27
The Emergence of Standards	30
Officials and Enforcement	34
3. THE TUDOR ERA	71
Changing Conditions	71
Early Tudor Standards	74
Tudor Officials	81
The Elizabethan Standards	86
4. EPILOGUE	94

APPENDIXES

A. *Weights and Measures of Merchandise in the English Import and Export Trade, ca. 1500 to ca. 1800*	103
B. *British Pre-Imperial Units*	141
C. *British Imperial Units*	161
D. *Pre-Metric Weights and Measures in Western and Eastern Europe*	169
BIBLIOGRAPHY	193
INDEX	235

Figures

1. English Linear Measures from 1497 to 1844 — 76
2. Exchequer Bronze Gallon and Bushel (1497) of Henry VII — 79
3. Avoirdupois Bell-Shaped and Flat Bronze Weights (1497) of Henry VII — 80
4. Exchequer Avoirdupois Bell-Shaped and Flat Gunmetal Weights, First Series (1558) of Elizabeth I — 86
5. Exchequer Avoirdupois Bell-Shaped Gunmetal Weights, Second Series (1574) of Elizabeth I — 89
6. Exchequer Troy Cup-Shaped Bronze Weights, Third Series (1588) of Elizabeth I — 90
7. Exchequer Avoirdupois Bell-Shaped and Flat Bronze Weights, Third Series (1588) of Elizabeth I — 91

Acknowledgments

H. Barrell and Julia B. Johnson of the National Physical Laboratory, Teddington, Middlesex, England, supplied me with pertinent information on, and several photographs of, current British imperial and metric physical standards. Dr. Barrell, a former Superintendent of the Standards Division, and Miss Johnson, of the Publicity Section, also helped me locate the sources for additional photographs, which I eventually received from John Moss, Windsor, Berkshire, a member of the American Society of Magazine Photographers, and from the Science Museum in London. A. F. Constantine, the Divisional Chief Technical Officer of the British Standards Institution, gave me several references to technical institutes that were able to supply additional data on the imperial and metric systems.

In the United States, three members of the National Bureau of Standards in Washington, D.C., A. G. McNish, L. J. Chisholm, and Ross L. Koeser, extended their resources and expertise to me repeatedly throughout the research stage of this history. Dr. McNish, Assistant to the Director, was especially helpful for his penetrating insights into current British and American efforts to adopt the metric system. Mr. Chisholm, a Special Assistant to the Manager of Engineering Standards, contributed some bibliographic aids, especially those concerning monographs, while Mr. Koeser, a Technical Researcher in the Office of Weights and Measures, sent me the recent Bureau of Standards publications on United States physical standards and standardization legislation.

Special recognition must be extended to Jon Eklund, Curator of Chemistry and Metrology, Museum of History and Technology, Smithsonian Institution, for his reading of the first draft and for his numerous insights and helpful suggestions.

Finally, I wish to thank the Committee on Research of Marquette University for several grants and a summer fellowship; Joan M. Krager, formerly General Editor of the University of Wisconsin Press, for unstinting encouragement and advisement; and Eleanor Woodward Sacks and John Berens for their professional secretarial competence.

R. E. Z.

Marquette University
Milwaukee, Wisconsin
December 1976

Introduction

Metrological development in the British Isles has undergone almost twenty centuries of evolutionary growth and decay. From the era of Caesar's exploratory expedition and Claudius' legionary colonization in the first centuries before and after Christ, through the period of the Anglo-Saxon, Danish-Norwegian, and Norman invasions of the Middle Ages, to the far-flung political and commercial empire of the modern age, the British peoples have produced one of the world's most comprehensive and sophisticated systems of weights and measures. Since the Anglo-Saxon period they have endeavored to simplify, refine, and standardize this system through the issuance of legislation, the construction of physical standards, and the setting up of inspection, verification, and enforcement procedures. Some of their attempts have been enormously profitable and others have proved totally futile. This book is a history of these successes and failures from antiquity to the end of the Tudor dynasty in 1603. The latter date serves as an appropriate and fitting climax because all of the developments described in this work constituted the foundation for the creation and implementation of the imperial system, which developed thereafter during the Age of Science, the Enlightenment, and the Industrial Revolution. The post-1603 developments will be the principal subjects of a volume to be issued later, which will continue the story of British metrological growth from the accession of the Stuarts to the present day.

The present work is an outgrowth of research conducted both before and after the publication of my first book, *A Dictionary of English Weights and Measures from Anglo-Saxon Times to the Nineteenth Century* (Madison: The University of Wisconsin Press, 1968), referred to in this work as "the *Dictionary.*" I am preparing a second edition of this dictionary which will extend its coverage from the early nineteenth century to the present.

As the bibliography in this book will testify, the literature of British weights and measures is both copious and multidisciplined. Not only are there many monographs and articles dealing specifically with weights and measures units and physical standards, but publications in many other fields—economic and social history, the history of science and technology, agriculture, medicine, pharmacology, philology, and law, to name a few—are concerned with one or more aspects of historical metrology.

There are, however, significant deficiencies in this literature, which the present work hopes to alleviate.

First of all, there is no publication of any kind that examines all three major components of British metrological history; that is, the units of measurement, the physical standards, and the corps of officials who inspected and verified weights and measures and who enforced compliance to the dictates of the law. In some instances, there is sufficient treatment of the units of measurement, their origins, and their change over time, but seldom are they related to their statutory definitions or to the physical standards constructed, ostensibly, as their models. Too often these units are treated in total isolation. Although some works discuss one or more of the physical standards, no attempt has ever been made to chart their continuous development from the Anglo-Saxon period. Weights and measures officials are almost never discussed, and if they are they receive only brief mention as a footnote to a presentation of the units. I have endeavored to give a proper perspective to all three of these components and to show how their interrelationship was responsible for the course that British metrology has taken over time.

Second, in examining any of these components, there has been a tendency merely to scan the rich corpus of statute law and to make assumptions without thoroughly analyzing the hundreds of decrees, ordinances, assizes, and statutes promulgated since the Anglo-Norman period. That body of law not only set the pace for metrological development but it mirrored many of the problems of society that prevented rapid solutions to the outstanding metrological impasses. In this book the law is carefully set forth and linked to a number of other factors that either forwarded or impeded metrological progress.

A third fault of the literature to date has been the forced linkages between ancient and contemporary systems of weights and measures on the one hand and the British system on the other. Noting only the fact that some British units had dimensions or specifications similar to those in other systems, too often writers have assumed, without any documentary or archaeological evidence, that British weights and measures must have been derived from earlier prototypes. Recognizing the fact that in some instances these linkages are valid, I have avoided drawing any analogies unless they are either explicitly made manifest in documents or are recognizable implicitly through commercial, industrial, social, agricultural, or political contacts and exchanges.

There has also existed a woefully inadequate treatment of the development of weights and measures in Scotland. The English system has too often been seen as completely divorced from the one employed by the Scots. I have tried to show the interconnections between Scottish and

English metrology, especially as they relate to the development of weights and measures law and the establishment of a permanent metrological officer corps. At least as early as the twelfth century, each of these two nations began to influence the course of many metrological developments within the other's borders until the Act of Union in 1707 created a common, united English system of weights and measures for both countries.

Finally, few previously published works have attempted to place any aspect of British metrology within a chronological, historical framework. By treating the various divisions of measurement in separate sections (e.g., one chapter on linear measures, a second on superficial measures, a third on weights, and the like), they have made it difficult, if not impossible, to demonstrate how developments in one of these divisions affected contemporary changes in the others. Too, such a format has almost precluded the possibility of connecting the significant changes in the development of weights and measures with those external forces that, traditionally, have interacted with any system of metrology, such as political influences, social upheavals, commercial contacts, and intellectual and scientific discoveries. Consequently, I have adhered to a chronological presentation by beginning with the Roman developments and ending with the contributions of the Tudors. In this way the history of British weights and measures becomes a truly integrated facet of British history.

There are four chapters in the text. Chapter 1 concentrates on the development of British weights and measures from the Roman to the end of the Norman period. Chapter 2 is concerned with the proliferation of weights and measures on the local and state levels during the High and Late Middle Ages and with the numerous attempts of medieval parliaments to achieve national uniformity. The monumental contributions of the Tudor councils, committees, and parliaments are discussed in the third chapter. The final chapter exhibits the long-range effects of these previous developments and describes the dominant metrological growth patterns after the accession of the Stuarts.

Following the text are four appendixes. The first is a list of the major import and export products of England from 1500 to 1800 and the statutory weights and measures by which they were bought and sold at port customs stations. The products were compiled from all of the available port records, merchants' manuals, books of rates and prices, and pertinent manuscripts and monographs. Appendixes B and C are tables of statutory pre-imperial and imperial units respectively. They are designed to illustrate which units were officially regulated and which were multiples of submultiples of others. Readers desiring more detailed and specific information on these state-regulated units or information on local units are referred to the *Dictionary*. The fourth appendix tabulates the principal pre-metric weights and measures units of western and eastern European countries during the nineteenth and twentieth centuries. I have presented them so that readers may compare and contrast them with their imperial and metric counterparts. Further correlations can be made by comparing them with the thousands of units found in the *Dictionary*.

A comprehensive, annotated bibliography and an index follow the appendixes.

BRITISH WEIGHTS & MEASURES

*A History
from Antiquity
to the Seventeenth Century*

1

Roman, Anglo-Saxon, and Norman Legacies

Roman Britain

From the first century A.D. until the Germanic invasions intensified in the latter half of the fourth century, Britain was tied to the mission and to the destiny of the Roman Empire. After the four legions under the command of Aulus Plautius landed in southern Kent in A.D. 43, the people of this island lived, toiled, and died under the Roman banner. Over the course of almost four hundred years the Catuvellauni, Trinovantes, Brigantes, and a host of other Celtic tribes acquired all of the material and spiritual benefits that the Rome of the Caesars felt obligated to bestow on cultures less advanced and less disciplined than its own. Representing the last significant conquest by the ancient world's most formidable military machine, this imperial province—so far removed from the Mediterranean nucleus of the Roman state—was stubbornly defended and lavishly subsidized by some of the most politically powerful and socially illustrious of Rome's citizens.

Britain was originally explored by Julius Caesar in the sixth decade before Christ, and its initial colonization was conceived and planned almost a century later by the Emperor Claudius. Julius Agricola, the eminent legion commander, became its sixth governor in A.D. 78 and held office long enough to see Roman domination extend into the Scottish highlands. Already, construction of the great roads such as Watling Street, Fosse Way, and Ermine Street had been completed under imperial auspices and the future urban centers of Lincoln, York, Gloucester, and Colchester had expanded beyond their military perimeters. Along the roads, lying adjacent to the cities, and strategically located in the Celtic countrysides, scores of forts and fortlets, called castra and castella respectively, dotted the landscape and served as an ominous reminder to the native inhabitants of the seriousness and the depth of the Roman resolve.

Owing to this military occupation and to the internal security that it generally afforded, merchants and tradesmen from many of the old centers of the empire rushed to this new land to tap what the Roman propaganda makers had labeled a collection of unlimited resources. Markets quickly arose and the products of British agriculture, mining operations, and small-scale crafts were shipped to France, Spain, Italy, and the southern and eastern provinces in exchange for Roman wines, pottery, grain, manufactured items, and many of the other necessities of life. But the fortresses and roads and veteran legions proved insufficient to maintain an effective hold on this province, and by the end of the first half of the second century two emperors—Hadrian and Antoninus Pius—had been forced to come to Britain to supervise, either directly or indirectly, the construction of two massive walls. Located about 130 kilometers from one another in northern England, these two gigantic cordons stretched from sea to sea, together consumed almost thirty years of the manpower and material resources of the province, and, temporarily at least, prevented the fiercely hostile northern tribesmen from joining with their southern kinsmen and disrupting the Pax Romana, however unpeaceful, in reality, it had always been.

Once again the Roman way reasserted itself and Britain was made secure in its political, economic, and social sectors. But, as before, the peace was short-lived. Teutonic inroads at the end of the second century and increasingly throughout the next two made it necessary for Rome to devote an even greater share of its already overly extended military force and of its depleted and depreciated currency to insure that Britain would not be lost. As on Rome's eastern front, reputations now depended on the degree of success or failure in stemming the rising tide of German expansion, and those involved in the defense of Britain were no exception. Commodus, Septimius Severus, Caracalla, Diocletian, Constantius Chlorus, Theodosius, and Honorius were only a few of the Roman leaders during this period who found it necessary either to alter the existing policy in Britain or to lead armies in its reconquest. But even during these years of turmoil, Roman fortitude proved equal to any challenge and Britain remained fertile ground for Roman industrial and commercial enterprise.

In the end, however, the goal proved unattainable. When the Visigothic army of Alaric threatened the very existence of the city of Rome at the beginning of the fifth century, Britain's legions were hastily withdrawn, never to return again. The island was freed at last from Roman domination but at a tremendous cost. As the empire declined and the western portion of it fell to a congerie of new and alien peoples, Britain acquired another conqueror, one who was equally militaristic but far less endowed in the arts of civilization. By the time this century had run its course, Britain was well on its way to becoming an Anglo-Saxon enclave while the rest of the Roman West was acquiring a Frankish, Visigothic, Vandal, and Ostrogothic character. One segment of British history was over; another was just beginning.

Since Rome's position in Britain had been insured only through a continuous military presence, it should not appear surprising that Roman soldiers were responsible for the introduction of Roman metrology and that for nearly 400 years these weights and measures were a permanent feature of British economic and social life. Primarily Greek and Egyptian in origin, Roman weights and measures had experienced a number of significant alterations during the pre-Christian, republican era. Thereafter, they were subject to very few changes and the system that Britain inherited was the best the ancient world had ever produced.

More important, like the Latin language and Roman law, this metrology demonstrated yet another feature of the universality of Roman life. Wherever one went in the empire the same measuring units and physical standards could be found. The pes, libra, and sextarius of the Eternal City were the same as those sent to Syrian, North African, Spanish, and British markets. The linear measures used by the Romans in the construction crafts, the superficial measures employed on the vast agricultural estates, the itinerary measures by which road and postal distances were calibrated, the capacity measures used by merchants and producers in the import and export trade, and the weights that formed the basis of Rome's monetary system were the same everywhere regardless of ethnic, cultural, geographic, and demographic differences. To insure the international uniformity of these weights and measures, Rome periodically sent copies of its physical standards to each province and enforced their proper maintenance and use. Not until the dissemination of metric weights and measures over most of Europe during the nineteenth century would such a condition exist again.

However, just as in Italy itself, provincial populations maintained indigenous systems of weights and measures. The people who cultivated farms or who tended their herds and flocks or who manufactured tools and other items on a small scale continued to employ weights and measures inherited from their ancestors. Many of these units survived the Roman period and evolved into the myriad of local systems so characteristic of pre-metric Europe. Britain was no exception. The Celts used their own weights and measures long before the Roman invasion and continued to use them long after the last of the legions departed. Even though no documentary evidence exists on the subject and no archaeological finds have ever provided us with any Celtic weights and measures, they must have been similar to those employed in most pre-industrial societies.

Undoubtedly, measurements were based on some simple or primary standards such as the pace, palm, or finger, or upon some arbitrarily defined standard such as a given number of grains. It is also safe to assume that each Celtic tribe had its own system of units, specially adapted to the particular needs of its people and to the resources found in any region. But regardless of how sparse or prolific these customary systems actually were, the Roman occupation had the profound effect of permanently localizing them. It was Roman metrology that was utilized whenever goods were exchanged in regional and interregional trade, whenever large construction projects were undertaken, whenever taxes had to be assessed on land and material possessions, and whenever coins had to be minted. And it was the Roman system of weights and measures that became the cornerstone for the future metrology of the British Isles.

Roman linear measurement was based upon the Roman standard foot (pes). This unit was divided either into 16 digits or into 12 inches. In both cases its length was the same. Metrologists have come to differing conclusions concerning its exact length but the currently accepted modern equivalents are 296 millimeters or 11.65 British imperial inches.[1] Expressed in terms of these equivalents, the digit (digitus), or 1/16 foot, was 18.5 millimeters (0.73 BI inches); the inch (uncia or pollicus), or 1/12 foot, was 24.67 millimeters (0.97 BI inches); and the palm (palmus), always defined as the length of ¼ foot, was 74 millimeters (2.91 BI inches). The Roman digit was nearly identical in length to the Egyptian digit of 1/54 meter, or 0.729 BI inches, while the Roman foot was slightly smaller than its Greek prototype, being 24/25 of the Greek foot of 308 millimeters (12.15 BI inches).

Larger linear units were always expressed in feet. The cubit (cubitum), for example, was 1½ feet (444 millimeters, or 17.48 BI inches). Like the Roman foot, it was smaller than its Greek prototype of 463 millimeters (18.225 BI inches). Five Roman feet made the pace (passus), equivalent to 1.48 meters, or 4.86 BI feet, which was usually divided into quarters (palmipes) of a foot and a palm. Ten feet made the perch (decempeda or pertica), which was equal to 2.96 meters (9.71 BI feet).

The most frequently used itinerary measures were the furlong or stade (stadium), the mile (mille passus), and the league (leuga). The stade consisted of 625 feet (185 meters, or 606.9 BI feet), or 125 paces, and was equal to 1/8 mile. During the republican era the Romans had borrowed the Greek stade of 600 Greek feet and rated it as equivalent to 625 Roman feet. The mile was 5000 feet (1480 meters, or 4856 BI feet), or 8 stades. It was similar to a Greek unit consisting of 8 Olympic stadia, and the

---

1. Hereafter the abbreviation "BI" will denote "British imperial."

Romans generally defined it for military purposes as the length of 1000 paces or double steps. The league had 7500 feet (2220 meters, or 7283 BI feet), or 1500 paces. For very large itinerary or topographical measurements, Roman engineers sometimes employed the schoenus of 4 miles, or 20,000 feet (5920 meters, or 19,424 BI feet).

On the latifundia of the Mediterranean world and on the provincial estates the fundamental superficial measure was the actus. It was 120 feet long and 4 feet (2 furrows) wide, and various combinations of this unit were arranged either horizontally or vertically depending on the type of agricultural operations being performed, the size of the work force, and the peculiarities of local topography. For instance, 30 such acti laid out in a horizontal pattern produced a square actus, or actus quadratus, a piece of land 120 feet by 120 feet and equal to approximately 50 BI square perches. This unit was ideally suited for small farming operations. A double actus quadratus made a jugerum, originally a day's work for a yoke of oxen. It was equal to 0.623 BI acre, or 28,800 BI square feet. When this most commonly used land parcel was doubled in size a bina jugera, or heredium, was produced. It was approximately 25 percent larger than Britain's current acre. The managers of large estates preferred to subdivide their fields into blocks consisting of 4 heredia arranged in a vertical line. This rendered a strip of land 480 feet long and 120 feet wide.

Prior to the third century B.C. the standard for all Roman weights was the as, or Old Etruscan or Oscan pound, of 4210 BI grains (272.81 grams). It was divided into 12 ounces of 351 BI grains (22.73 grams) each. In 268 B.C. a new standard was created when a silver denarius was struck to a weight of 70.5 BI grains (4.57 grams). Six of these denarii, or "pennyweights," were reckoned to the ounce (uncia) of 423 BI grains (27.41 grams) and 72 of them made the new pound (libra) of 12 ounces, or 5076 BI grains (328.9 grams).[2] This pound was brought to Britain and became the standard there, as well as in the rest of the empire, for weighing gold and silver and for use in all commercial transactions.

For mercantile purposes, this new pound had three subdivisions. It consisted of 12 ounces, each ounce containing 24 scruples, and each

---

2. The weight cited for the silver denarius constitutes the mean of a number of extant denarii and is currently the most generally accepted figure. The weights for the ounce and pound are in direct correlation to this mean. In the past, two other metric and British imperial equivalents were common: A. E. Berriman in *Historical Metrology* (London, 1953), p. 27, assigned a weight to the post-268 B.C. pound of 5040 BI grains (326 grams) but did not state how this figure was derived; and Edward Nicholson in *Men and Measures: A History of Weights and Measures: Ancient and Modern* (London, 1912), pp. 40-41, recorded a much larger figure—arguing that the weight of the new pound was related to a percentage taken from the Alexandrian talent, Nicholson computed the pound at 5244 BI grains (338 grams).

scruple (scrupulum) containing 2 obols. In the mints the pound again consisted of 12 ounces, but the ounce was divided into 6 sextulae, and the sextula or nomisma into 24 siliquae.

The principal Roman capacity measures were the hemina, sextarius, modius, and amphora for dry products, and the quartarus, sextarius, congius, urna, and amphora for liquids. As with linear measures and weights, there is no general agreement among metrologists as to their metric or British imperial equivalents. Since Roman writers did not define capacity measures in terms of cubic inches but rather as fractions or multiples of the sextarius, it is necessary to rely completely on the calibration of any given number of extant standards. Since no two standards are ever identical, only an approximate cubic capacity can be determined. In the case of the sextarius, the mean generally agreed upon today is 35.4 BI cubic inches, or nearly 1 BI pint (0.58 liters). The hemina, or half-sextarius, based on this mean was 17.7 BI cubic inches (0.29 liters). Sixteen of these sextarii made the modius of 566.4 BI cubic inches (9.28 liters), and 48 of them made the amphora of 1699.2 BI cubic inches (27.84 liters).

In the liquid series, the quartarus, or ¼ of a sextarius (35.4 BI cubic inches), was 8.85 BI cubic inches (0.145 liters). Six of these sextarii made the congius of 212.4 BI cubic inches (3.48 liters), 24 sextarii made the urna of 849.6 BI cubic inches (13.92 liters), and, as in dry products, 48 sextarii were equal to one amphora.

Anglo-Saxon Britain

These weights and measures, and undoubtedly others never mentioned in any of the sources, were used in Britain during the occupation. They did not suddenly disappear when the legions and civil bureaucracy returned to Rome, but the unstable political conditions of the Early Middle Ages and the fundamental differences between Roman and German institutions profoundly altered both their importance and application.

Roman institutions had evolved to meet the needs of a very complex, heterogeneous civilization. Stretching from the Atlantic Ocean into the Slavic portions of eastern Europe and from the German North to central Africa, the empire conducted its commercial, industrial, and agricultural pursuits over immense distances and among hundreds of divergent socioeconomic groups. To maintain a population of almost 100,000,000 by the second century, Rome depended on the rapid transport of grain and other food staples over many sea and land routes and on the massive warehousing of foodstuffs to combat the ravages of periodic famines and plagues. Its efforts in establishing an efficient postal service, a comprehensive bureaucratic network, a sophisticated legal and judicial system, and the ancient world's largest welfare program were also made imperative by the empire's geographic and demographic dimensions. Roman metrology—essential to this economic machine—merely reflected the variety and the complexity of this experience.

In Britain these weights and measures declined in importance and either ceased to be employed or evolved along different lines after the fourth century because Anglo-Saxon institutions could not sustain them. In their former homeland along the North and Baltic seas, as well as in their new British settlements, these people—when not engaged in war under local comitatus leaders—farmed small parcels of land in village hamlets and watched over their flocks in nearby fields. Their crafts were primitive compared to those of the Romans, and their trade—always local in nature—consisted principally in the exchange of foodstuffs and raw materials. They had no written, codified law and no professional class of judges. The witans—councils of wisemen who advised tribal chiefs in matters concerning war, peace, and marriage alliances—were their only assemblies. This was a warrior society, one that was distinguished by its constant mobility, by its perpetual military activity, and by its never-ending search for new land. Because of their overriding concern for migration and war, they were by nature politically unstable and economically underdeveloped.

But even if the Anglo-Saxons had possessed political, legal, or economic equality with their Mediterranean neighbors, they would have found it impossible to maintain intact that system of metrology inherited from the Romans because of the radically different political, economic, and social conditions of the Early Middle Ages. Until a temporary unification of Britain was achieved in the early ninth century, conditions there, as well as in the rest of western Europe, were abnormally chaotic and violent. Early medieval man witnessed the complete breakdown of Roman solidarity and the embryonic beginnings of feudalism. In place of an internationally cohesive Roman government, there now existed a potpourri of Germanic kingdoms, each of them exercising political and military power over a very small area. Europe was evolving from a southern Latin orientation to a northern German one. Isolation and seclusion became the principal ingredients of the world north of the Alps. Merchant caravans virtually ceased traveling over the commercial highways since protection could no longer be afforded them. Regional and interregional trade dwindled and the urban centers lost their most energetic and dynamic elements. Local affiliations, temporarily stultified under Roman rule, reasserted themselves. The industrial, commercial, and monetary empire of Rome gave way to a landlocked, agricultural environment in which local exchange in kind and services sufficed for most people's needs. Consequently, Roman metrology—like most other aspects of Roman civilization—was far too sophisticated and advanced for the demands of this new age.

Everywhere in western Europe the attention of each newly arrived Germanic tribe was focused by necessity on permanent settlement. In Britain, from the beginning of the fifth century to the end of the sixth, the kingdoms of Northumbria, Mercia, East Anglia, Wessex, Sussex, Essex, and Kent became political realities, but only after their kings devoted all of their energies to crushing the last remnants of Romano-Celtic resistance. The reputations of Hengest, Horsa, Cerdic, Cynric, Aelle, Ceawlin, Ethelbert, and others were earned from their successful campaigns against the non-Germanic inhabitants and not from their occasional attempts to better the economic, legal, and social life of the people.

Once settlement was achieved, these autonomous kingdoms warred uninterruptedly against one another. In the seventh century the military prowess of Edwin, Oswald, and Oswiu enabled Northumbria to extend its influence to almost all corners of Britain, while in the eighth Ethelbert and Offa of Mercia made that kingdom supreme. Not until the fortunes of Wessex rose in the ninth century under Egbert and Alfred was there any attempt to fashion a government, a law, and an economy even remotely similar to those that had existed in Roman days. And no sooner had Wessex provided conditions for a more permanent and beneficent society than the island was again overrun, this time by Scandinavian pirates. Domestic turmoil was reborn and yet another foreign dimension was added to the British experience.

Like the Celts, the Germanic tribes—both on the continent and in Britain—used weights and measures that were based on simple, primary standards. The North German foot—the standard for linear measurement—was defined as the length of 12 thumbs, or of 36 barleycorns laid end to end. All other linear measures were expressed in fractions or multiples of this foot. The palm, consisting of 3 thumbs, or 9 barleycorns, was ¼ foot. For building purposes there was a cubit of 2 feet. Four feet, or 2 cubits, made the cloth-elne, the customary width of woolen and other cloths woven on the handlooms of the period, and 15 of these feet made the rod, used for land surveying and agricultural measurements.

Considerably larger than its Roman prototype, the North German foot has been calibrated at 13.2 BI inches, or 335 millimeters. Based on these equivalents, the thumb of $^1\!/_{12}$ foot was 1.1 BI inches (27.9 millimeters); the palm, or ¼ foot, was 3.3 BI inches (83.8 millimeters); the cubit was 26.4 BI inches (670 millimeters); the elne was 52.8 BI inches, or 4.4 BI feet (1.34 meters); and the rod was 16.5 BI feet (5.03 meters). Both the Roman and German feet were used in Britain during the Early Middle Ages but by the end of the period, or prior to the Norman Conquest, the German foot had emerged as the primary unit and would become the standard for all English linear measures after 1066.

One of the principal reasons for the eventual ouster of the Roman foot was that the Anglo-Saxons continued to employ their traditional agricultural field divisions after their arrival in Britain, and superficial measures have always depended on the standard of linear measurement. In fact, the Roman foot was relegated almost immediately to the construction crafts while the German foot became the standard for all other pursuits. In plowing, Saxon peasants worked a section of land that consisted of 40 rods in length (the furlong) and 4 rods in width. This parcel of 160 square rods, or 36,000 square feet, was the acre. It was exactly the same size as the modern English acre of 160 square rods, or 43,560 square feet, since the northern rod of 15 German feet was the same length as the modern rod of 16.5 BI feet.

Before the Norman Conquest the standard for weights was the Saxon or moneyer's pound. Its earliest regulation occurred during the reign of Offa (757-96), who instituted a new coinage to replace the erratic currency of the Mercian realm. He established a silver penny (called a sterlingus, sterling, or esterlingus) of 22.5 BI grains, which set the standard of silver coinage in England until 1344 and of the English weight system for gold, silver, jewels, and electuaries (meaning either an alloy of gold and silver, or amber) until 1527. Twenty of these pennies by weight (called pennyweights) made the ounce of 450 BI grains, and 12 such ounces made the pound of 5400 BI grains.[3] The grain content of this pound was the same as the Arabic silver rotl. Until the heavier commercial pounds appeared after the conquest, this Saxon pound served all monetary and commercial needs in the same manner as the pre- and post-268 B.C. Roman pounds of antiquity.

There is some evidence for the existence of other Anglo-Saxon weights and measures. In a decree issued sometime during the seventh or eighth decade of the tenth century, Edgar the Peaceable ruled that "the measure of Winchester should be the standard for his realm" (*et una mensura, sicut apud Wincestram*). Since Winchester was the capital for Edgar and other Anglo-Saxon kings after the Heptarchy and the Bretwalda system

---

3. All of these weights are expressed in troy grains. Actually the Saxon pound—called the tower pound after the conquest—contained 7680 wheat grains, which are converted to the troy barleycorn scale as follows: 32 wheat grains = 22.5 barleycorns. Therefore, (1) 32 x 20 pennyweights x 12 ounces = 7680 wheat grains; or, (2) 22.5 x 20 pennyweights x 12 ounces = 5400 troy grains. The Offa penny thus weighed 32 wheat grains and the ounce of 20 pennyweights was 640 wheat grains.

had been replaced by a territorially unified, if not politically solidified, English state in the early ninth century, it is perfectly reasonable that its metrological system should be considered as the model for the English domain. It is also a well-known fact that Winchester housed the Saxon standards.[4] But was Edgar referring to the measure of Winchester as a system of units of measurement or as a particular physical standard or set of standards? G. T. McCaw, using some of the researches of H. W. Chisholm, a former director of the British Standards Institution, wrote that the yard or gird of the Saxon kings was kept at Winchester and he indirectly implied that this was the measure referred to in Edgar's statement. McCaw also remarked that there were other standards deposited at Winchester as early as the reign of Alfred the Great, or about the year 850.[5] John O'Keefe stated that Edgar kept his standards, which included a corn bushel, at Winchester, and O'Keefe inferred that the king was thinking of these when he issued his decree.[6] Unfortunately, Edgar's statement mentions no weights or measures by name, a characteristic of the wording in many of the laws until as late as the end of the Angevin period. There is no corroborating evidence in any Anglo-Saxon document that would prove what physical standard Edgar had in mind. No document ever says that the Crown standards were duplicated and distributed throughout the land. As a general rule, the documents merely plead for the use of just weights and measures, and they threaten lawbreakers either with secular or divine punishment.[7] Consequently, Edgar must have used the words "measure of Winchester" in their generic sense, implying units

    4. See "Seventh Annual Report of the Warden of the Standards on the Proceedings and Business of the Standard Weights and Measures Department of the Board of Trade for 1872-73," in *Parliamentary Papers,* Great Britain, *Reports from Commissioners,* 38 (London, 1873): 47; W. H. Prior, "Notes on the Weights and Measures of Medieval England," *Bulletin du Cange,* 1 (1924): 82–84; John Quincy Adams, *Report of the Secretary of State upon Weights and Measures Prepared in Obedience to a Resolution of the House of Representatives of the Fourteenth of December, 1819* (Washington, D.C., 1821), H.R. Document 109, p. 69; and John A. O'Keefe, *The Law of Weights and Measures* (London, 1966), pp. 2–3.
    5. G. T. McCaw, "Linear Units Old and New," *Empire Survey Review,* 5 (1939-40): 254.
    6. O'Keefe, *Law of Weights and Measures,* pp. 7-8.
    7. Two selected passages from the *Ancient Laws and Institutes of England* (London, 1840), p. 313, will demonstrate this: "Ut Nullus Injustas Mensuras, et Pondera Injusta, Lucri Causa Dare Praesumat" ["He presumes that no unjust measures and unjust weights will result in profit"]; and "Ut mensurae et pondera justa fiant, sicut in divinus legibus sancitum est; ergo statuimus ab omnibus hoc observandum" ["Just as it is confirmed in divine law that measures and weights be just, we establish, therefore, that this be observed by everyone"].

of measurement common to Winchester rather than their narrower, but more specific, meaning of physical standard.

As accurately as can be determined, none of the governmental and very few of the local Saxon standards have survived. The former, customarily called "pondi regis" and "mensurae domini regis" in early medieval sources, usually were melted down to produce new models. There is some documentary evidence, however, pointing to the existence of one particular state standard: the yard (gird or girda) preserved at Winchester for an unspecified period prior to the Norman invasion.[8] Since it has never been found and since no actual description of its physical appearance was ever made, one can only assume, judging by linear standards of a later medieval period, that it was an elongated, four-sided, lead or iron bar that indicated a specific length in one of two ways. It was either one yard long from end to end or it was slightly larger with a line marked near each end, the space between the two lines representing one yard. Even though the practice of stamping standards with the seal or mark of the monarch is not documented before the reign of William I, it would not be unlikely that some kind of seal was placed on this standard. Until some new material sheds additional light on this subject, nothing more substantive than the above can be claimed.

There is little evidence for the existence of other types of Saxon standards. Bruno Kisch has shown that in the Streeter Collection at Yale University there is an eighth- or ninth-century weight which is made of lead and covered with an ornamental surface of gilt brass. A standard made of more than one metal is indeed rare in English history. He also reports that, probably due to Roman influence (accruing either through Roman trade or the occupation itself), disk weights were used in England during the pre-Norman period in addition to the traditionally English bell-shaped weights.[9] It may be speculated that the weights referred to by Kisch were fractions and/or multiples of the Saxon pound. F. G. Skinner mentions that weights on a 48-grain basis for series of 3, 5, 6, and 12 units (that is, from 144 up to 576 grains) have been found in the graves of two moneychangers in Kent. These discoveries, at Ozingell in Thanet and at Grove Ferry near Canterbury, supposedly date from approximately the beginning of the seventh century.[10] Although no interconnecting link can

---

8. There are brief discussions of this particular standard in H. J. Chaney, *Our Weights and Measures: A Practical Treatise on the Standard Weights and Measures in Use in the British Empire with Some Account of the Metric System* (London, 1897), pp. 23-24, and in Prior, "Weights and Measures of Medieval England," pp. 142-43.

9. Bruno Kisch, *Scales and Weights: A Historical Outline* (New Haven, 1965), p. 101.

10. Frederick George Skinner, *Weights and Measures: Their Ancient Origins and Their Development in Great Britain up to AD 1855* (London, 1967), p. 10.

be proven, it is perhaps more than coincidental that one of two Arabic dirhem standards of the ninth century weighed 48 grains, as did a much earlier Greek Aeginetan standard—the latter generally considered the major influence on the Arabic trade dirhem.

Aside from coinage standards, nothing else is extant. Numerous writers in the medieval and early modern periods, as well as a few in our own day, have attributed an original bushel to King Edgar. This famous monarch supposedly ordered its construction and then stored it along with "the other Anglo-Saxon standards" at Winchester. There has never been either documentary or archaeological proof of its existence. Its name never appears in any Anglo-Saxon manuscripts. Nor does the name of any other capacity measure. Besides, the etymology of "bushel" suggests an Old French or perhaps even a Norman origin and it is highly unlikely that a measure so named would have been so commonplace as to suggest a state standard before the advent of the Normans. Once again, until we learn a great deal more through finds of additional manuscripts or through significant archaeological discoveries, our knowledge of Anglo-Saxon standards will not improve appreciably.

Norman Britain

Regardless of how little is actually known concerning these units and physical standards, it is an indisputable fact that the Norman Conquest did not alter either their development or application in any way. The weights and measures of Winchester survived the conquest undisturbed and, together with those surviving from Roman times, were adopted by the new Norman overlords. There were two principal reasons for this. First, the Normans were vastly outnumbered in their new land and in order to minimize Anglo-Saxon hostility, they perpetuated any institutions and customs that did not interfere with the proper functioning of their new government. William replaced the old Anglo-Saxon witan with his own court, the curia regis, and staffed it with those Norman lords who had fought at the Battle of Hastings and with those members of the Anglo-Saxon aristocracy who had aided the Norman cause. Further, he seized all of England on behalf of the Normans and after retaining approximately one sixth of it for himself, he parceled out the rest to his curia members in the form of fiefs. These new barons or tenants-in-chief were prohibited from building castles without William's permission, were forbidden to engage in local wars without royal favor, and were encouraged to subinfeudate their lands in order to maintain a feudal army of approximately 5000 members. Aside from these programs and his strenuous effort to replace the Anglo-Saxon church hierarchy with Norman personnel, he did not tamper with any other Anglo-Saxon institution. The political and military life of the island was revolutionized; its commercial and agricultural life remained unchanged.

Second, there was no need to destroy or to amend most Anglo-Saxon institutions. The Normans were distinguished by their tolerance and by their ability to rule foreign areas; they honored the customs and traditions of subject peoples. In Sicily, Apulia, and Calabria, for example, they ruled populations that were far more multilingual and diversified than that of England and they provided the best government in those areas since Roman times. The Normans never interfered with effective or successful institutions. In Britain they merely added a further dimension to most institutions and either improved them or helped preserve them by ignoring them.

William's policy toward Anglo-Saxon metrology was in line with these general observations. Once the invasion and the sustained land war had succeeded in subduing Anglo-Saxon resistance to the Norman military, William concentrated on his domestic program. In one of his earliest decrees—the only one during his entire reign dealing with weights and measures—he reinforced the decrees of the Wessex kings that aimed at metrological uniformity.[11] In commanding that all weights and measures throughout the realm be uniform and stamped with his seal to authenticate them, he took as his model the Winchester system of his Anglo-Saxon predecessors. He then transferred the Winchester standards to London and deposited them, along with the royal treasures, in the Crypt Chapel (later called the Pyx Chapel or the Chamber of the Pyx) of Edward the Confessor, in Westminster Abbey.[12] William's motive in these maneuvers is quite apparent: he wanted to appear not merely as a military conqueror but as a lawful successor to Edward the Confessor. In the process, the weights and measures used by the Anglo-Saxons became the established metrology of Norman England.

---

11. William's decree on weights and measures may be found in Wilfrid Airy, "On the Origin of the British Measures of Capacity, Weight and Length," *Minutes of Proceedings of the Institution of Civil Engineers*, 175 (1909): 176, and in any of the statute books listed under Documents in the Bibliography at the end of this book. As with William, it occasionally has been popular to credit Canute (1016-35) with major advances in weights and measures reform. For example, Malachy Postlethwayt in *The Universal Dictionary of Trade and Commerce*, 2 vols. (London, 1755), 2:186, says that "there were good laws for weights and measures made before the conquest by Canute." John Quincy Adams in his *Report to the House of Representatives*, p. 22, says much the same thing. There is no evidence whatsoever for this acclaim. Canute merely carried on programs that were already put into operation by his predecessors, and he did not introduce any innovative or reformative legislation on the subject.

12. The Pyx Chapel was also the home of the standard trial plates for gold and silver coins used at the trials of the Pyx—the formal official assays of the coins of the realm.

2

Medieval Times:
Refinement and Proliferation

Angevin-Plantagenet Legislation

The decree of William ended the first phase of British metrological history. During that turbulent millennium of invasion, conquest, and settlement, Britain inherited the Roman and North German systems of weights and measures. By the Norman period the Germanic system had become dominant in the government's monetary, commercial, and agricultural operations, while in the recesses of the countrysides, in the mountainous regions, and on the channel and sea islands the indigenous Celtic system lingered on. Those units and physical standards that made up the Winchester system proved adequate for the needs of Norman society. Thereafter, economic and social conditions changed rapidly and a new era in British weights and measures was inaugurated.

This second phase spans the years during which the Angevin-Plantagenet and Lancastrian-Yorkist dynasties ruled England. Soon after the beginning of the thirteenth century the government embarked on a new metrological policy by initiating a comprehensive legislative program aimed at the establishment of a unified system of weights and measures. Accomplishing this goal of unit standardization was a formidable task but it was made even more difficult because of the dynamic changes that swept over most areas of western Europe during the High and Late Middle Ages. Specifically, regional and interregional trade had resurfaced. In Britain, commercial relations with Scotland, France, Flanders, the German Hansa, and Italy led to the borrowing of many foreign units, principally for the purpose of simplifying commercial and financial exchanges. New items of manufacture and significant increases in the imports and exports of such products as unprocessed raw materials, naval stores, leather goods, textiles, chemicals, wines, spices, and precious metals produced a great influx of new weights and measures. In turn, these units were influenced in varying degrees by faster methods of transport and distribution; by expanded industrial, mechanical, and technological processes; by revolutionary agricultural and mining techniques; and by the slow but persistent urbanization of what formerly had been a predominately rural domain. To limit their growth and to create a uniform system applicable to all sectors of the economy were the goals of pre-Tudor legislation.

To help achieve these goals, the Angevins ordered the construction of physical standards and had them disseminated throughout the realm; and they delegated the metrological powers of inspection, verification, and enforcement to people from many different levels of society. During William's reign the Winchester standards had been transferred to London but they were never duplicated and distributed to important markets and urban centers. The Angevin governments recognized the fact that the promulgation of a law—whether it be in the form of a decree, charter, ordinance, or statute—did not guarantee metrological uniformity. Needed were physical models of the state-regulated units so that officials in each region could conduct periodic inspections of weights and measures and verify either that local producers and merchants were abiding by the dictates of the law or that they were guilty of breaking that law.

Unit standardization was the aim of all of the early Angevins but it was not until the reign of King John (1199-1216) that any formal program was set into motion. It is surprising that so slow a start was made, since the first king of this new dynasty, Henry of Anjou, earned the sobriquet Lion of Justice for his pioneer work in helping create the English judicial system. This included perfecting the scope and functions of assizes, but Henry II (1154-89) and Richard I (1189-99) used assizes only to promulgate the issuance of physical standards and to regulate inspection, verification, and enforcement procedures; never to formulate governmental specifications for individual units.[1] This is true even of Richard's

1. In the Norman and early Angevin periods an assize either was an enactment that regulated the quality, quantity, weight, measure, and price of articles for sale or it was a session at which the examination and authentication of weights and measures took place. By the thirteenth century, however, the word "assize" retained generally only its first meaning; the latter process became known as an "assay." See *A Consuetudinary of the Fourteenth Century for the Refectory of the House of S. Swithun in Winchester,* ed. George William Kitchin (London, 1886), p. 78; White Kennett, *Parochial Antiquities Attempted in the History of Ambrosden, Burcester, and Other Adjacent Parts in the Counties of Oxford and Bucks* (Oxford, 1695), glossary, sv *assisa;* and David Macpherson, *Annals of Commerce, Manufactures, Fisheries, and Navigation,* 4 vols. (London, 1805), 1:357. In addition, various sections in the following works contain information on the historical development of assizes: William Stubbs, *Constitutional History* (Oxford, 1906); F. Kern, *Kingship and Law in the Middle Ages* (Oxford, 1956); E. Lipson, *The Economic History of England,* Vol. 1 (London, 1949); Edward Coke, *The Fourth Part of the Institutes of the Laws of England concerning the Jurisdiction of the Courts* (London, 1797); K. M. Murray, *A Constitutional History of the*

famous Assize of Measures (*Assisa mensurarum*) of 1197, which called merely for uniformity in all English liquid and dry measures without ever describing, or even mentioning, any by name.[2] Richard's assize probably resulted from the fact that his decree of 1189, demanding "one weight and one measure" in all commercial transactions, had failed completely in attaining its objective.[3] The assize of 1197 would suffer a similar fate.

The ambiguous phrase "one weight and one measure" in Richard's decree formed an essential part of chapter 35 of King John's Magna Carta in 1215. This section declared that throughout the kingdom one measure should be used for wine, one for ale, and still another for corn. In the case of wine and ale it never named the capacity measures, but it stated that the London quarter (the Saxon seam or pack-load) was to be the model for corn. It also established a uniform breadth for three types of fabrics—dyed cloths, russets, and haberjets—saying that they were to be two "ulne" within the lists.[4] The ulna was the Latin equivalent of the Saxon

---

Cinque Ports (Manchester, 1935); *British Borough Charters: 1216-1307,* ed. A. Ballard and J. Tait (Cambridge, 1923); *Somersetshire Pleas—from the Rolls of Itinerant Justices,* ed. B. Healy, Somerset Records Society (London, 1897); *Calendar of the Plea and Memoranda Rolls,* ed. A. H. Thomas (Cambridge, 1929); *Chronicle of the Mayors and Sheriffs of London,* ed. H. T. Riley (London, 1863); *Records of Some Salford Portmoots,* ed. J. Tait, Chetham Society, New Series, Vol. 80 (Manchester, 1926); *Selected Rolls of the Chester City Courts,* ed. A. Hopkins, Chetham Society, Third Series, Vol. 2 (Manchester, 1950); Cornelius Brown, *A History of Newark on Trent* (Newark, 1904); *Records of the City of Norwich,* ed. W. Hudson and J. C. Tingey (Norwich, 1906); *Court Rolls of the Manor of Ingoldmells,* ed. W. C. Massingberd (London, 1902); *Calendar of the Letter Books of the City of London,* ed. R. H. Sharpe (London, 1899); James Tait, *The Medieval English Borough* (Manchester, 1936); and J. E. A. Jolliffe, *Constitutional History of Medieval England* (London, 1954).

2. Matthaei Parisiensis (Matthew of Paris), *Historia Anglorum,* ed. Sir Frederic Madden, 6 vols. (London, 1865-69), 2:65, wrote: "Eodem anno constitutum est a rege Ricardo . . . ut omnes mensuræ per totam Angliam in bladis et leguminibus, tam in civitatibus quam extra, ejusdem sint quantitatis, et maxime mensura celiæ, vini, et pondera mercatorum" ["In this year it is decreed by King Richard . . . that all measures for cereals and legumes throughout England, within cities as well as without, be of uniform dimensions for oil and wine measures and for merchants' weights"]. See also ibid., 3:216; "Annales de Burton (A.D. 1004-1263)," in *Annales monastici,* ed. Henry Richards Luard (London, 1864), pp. 192-93; Macpherson, *Annals of Commerce,* 1:357; W. H. Prior, "Notes on the Weights and Measures of Medieval England," *Bulletin du Cange,* 1 (1924): 83; and William Stubbs, *Historical Introduction to the Rolls Series* (London, 1902), p. 299.

3. See *Chronica Johannis de Oxenedes,* ed. Sir Henry Ellis (London, 1859), p. 72; *Flores Historiarum,* ed. Henry Richards Luard, 2 vols. (London, 1890), 2:109; Bartholomaei de Cotton, *Historia Anglicana (A.D. 449-1298),* ed. Henry Richards Luard (London, 1859), p. 83; Matthaei Parisiensis (Matthew of Paris), *Chronica Majora,* ed. Henry Richards Luard, 6 vols. (London, 1874), 2:351, and *Historia Anglorum,* 2:10.

4. According to *Webster's Third New International Dictionary* (unabridged; Springfield, Mass., 1968), a list is the border, strip, or selvage on either side of a woven or flat-knitted fabric so finished as to prevent raveling. Sometimes it is woven of different or heavier threads than the fabric and sometimes in a different weave.

elne, or "elbow," hence "forearm length." The chapter ending insisted on uniformity for weights as well as for measures.[5]

The charter is unique only to the extent that it is the first Angevin enactment that specifies a particular unit—the London quarter. Although it did not define its size, contemporary documents frequently refer to the quarter as a measure consisting of 8 bushels, the latter probably originating as a northern French or Norman measure. Since the prices of bread and, to a lesser degree, of ale depended on that of corn, the importance of this clause was very great. The problem of ambiguity inherited from Richard's laws remained, however. The framers of this chapter refer specifically and exclusively to measures for three articles but then only promulgate a unit standard for one of them. Since they did not feel obliged to mention the other two, they obviously were interested only in attaining a uniformity of proportion among the three measures and not one of identity. They hoped to fix existing rights and usages as a defense against fraud and oppression. They wanted to insure that the measures for wine, ale, and corn would never be identical.

The failure of this chapter can be seen in the fact that it was part of a reissue of Magna Carta by Henry III in 1225 (chapter 25 in this document) and would reappear in more than two dozen statutes until the reign of Charles II. Frequent repetition meant noncompliance. In the Magna Carta of 1216 for Ireland the same wording is used except that the weights and measures of Dublin were declared the standards. During the reign of Henry III, however, the viceroy and council ordered observance of the London standards.[6]

The long reign of Henry III (1216-72) was somewhat of a turning point in the history of unit standardization. In addition to promulgating one decree in 1216 that reconfirmed all of the laws of his predecessors, a second in 1228 that ordered compliance to the weights and measures provisions of Magna Carta, and a third in 1233 that prohibited the measurement of cloth by the "yard and handful,"[7] Henry III in 1266 was partly responsible for the creation of the famous Assize of Bread and Ale (*Assisa panis et cervisiae*). This assize regulated the weights that applied

---

5. The following wording of chapter 35 may be found in any of the statute books listed under Documents in the Bibliography at the end of this book: "Una mensura vini sit per totum regnum nostrum et una mensura cervisie et una mensura bladi scilicet quarterium London' et una latitudo pannorum tinctorum russettorum et halbergettorum scilicet due ulne infra listas. De ponderibus autem sit ut de mensuris."

6. See *Historic and Municipal Documents of Ireland, A.D. 1172-1320,* ed. J. T. Gilbert (London, 1870), pp. xxxiv, 70.

7. In London it was customary for a draper in measuring cloth to mark the end of each yard by placing his hand on the cloth and starting the next yard from the other side of the hand. In this way he gained 2 yards every time he sold 24. The prohibition was not very effective since this practice was still common during the reign of Elizabeth I.

to bread, established certain maximum prices for bread and ale, and ordered various fines and forms of punishment for bakers and brewers convicted of breaking the assize.[8] It declared that bread must be weighed according to the king's standard pound, which, based on the description in the document, was the tower pound of 7680 wheat grains, which equaled 5400 troy grains or barleycorns. This was the first time that Offa's Saxon pound system was recognized officially through a state enactment.

This assize was important for more than regulation of the sale of bread and ale. Henry issued it for political and commercial reasons. Just prior to its passage he had granted to the merchants of Hamburg and Lübeck permission to establish their own Hansa in London. The Gotland Association, including the merchants of Cologne, had been given similar privileges as early as 1237. As in England, merchants in all of these northern German and Baltic cities used the 450-grain ounce of the Arabic silver dirhem standard for their coinage and commercial weights. An 8-ounce mark of 3600 BI grains (233.28 grams) was used in the mints and a 16-ounce pfund of 7200 BI grains (466.56 grams) served all mercantile needs. The northern German commercial pound was one ounce (450 grains) heavier than the commercial pound of southern Germany and of Scotland, which totaled 6750 grains. The latter pound, also used in England during this period, consisted of 15 ounces of 450 grains each. Henry's action was timed perfectly in order to establish some common basis of exchange with the Hansa merchants, and this was extremely important in an age when the German towns were fast becoming the principal importers of English wool.

The last important document of the thirteenth century, the Composition of Yards and Perches (*Compositio ulnarum et perticarum*), can be attributed either to Henry III or his successor, Edward I (1272-1307). Its exact date is not known and its authorship is still in doubt. It must, however, have been drawn up after the Assize of Bread and Ale of 1266 and before the Treatise on Weights and Measures (*Tractatus de ponderibus et mensuris*) of Edward I in 1303. No source before 1266 ever mentions the *Compositio,* and its formula for linear and superficial measurement, along with the formula for the construction of the tower pound found in Henry's assize, are repeated word for word in Edward's *Tractatus.* It could be argued that Edward issued the *Compositio* merely to regulate linear and superficial measures and later decided to broaden its scope by including the tower pound. Hence he combined his *Compositio* with Henry's formula and produced a more representative treatment of

---

8. The fines depended on the degree of the offense and the number of previous convictions, but frequent offenders were subjected to the tortures of flogging, the pillory, or the tumbrel.

contemporary units. An argument that can be made on Henry's behalf is that since he already had promulgated standard units of weight, and since he believed that his assize was being effectively received and enforced, he issued another document to cover measures of length and area. Also, since Edward had probably copied verbatim Henry's section of the assize dealing with the tower pound, he would be likely to do the same with Henry's *Compositio*. Neither of these two interpretations can be proven conclusively. The most anyone can say with certitude is that the document was written between 1266 and 1303 and that it was the first governmental enactment in England regulating measures of length and area. The following list presents its essential features.

3 round and dry barleycorns[9]	=	1 inch
12 inches	=	1 foot
3 feet	=	1 yard
5½ yards	=	1 rod (perch)
40 rods in length and 4 in breadth	=	1 acre

This ordinance is immensely important because it represents a complete break with earlier Anglo-Saxon tradition. All linear measures were now based on a standard yard, or "iron ulna," which was 3 feet in length, each foot, as in Roman metrology, divided into 12 inches. There were other ulnas in use, particularly the old Saxon elne employed as a basis of cloth measurement in Magna Carta. They were eliminated as standards and the newly constructed yard became the primary standard on which all other linear measures depended or were referred to for comparison throughout the Middle Ages. The original iron measure has disappeared but it is unlikely that it differed from the present standard by more than 0.04 BI inches (1 millimeter) as shown by the gauging in this century of several of its derivatives, including the Exchequer standards of Henry VII and Elizabeth I.

The *Compositio* was a compromise. Its purpose was to establish uniformity in view of the diversity of foot units, principally Roman and North German, then in use. But it was an illogical compromise. The obvious course would have been to construct a new yard standard consisting of 3 Northern feet. In this way 5 yards, or 15 feet, would have made a rod instead of the very awkward factors of 5½ yards, or 16½ feet, created in order to retain dimensions for the acre exactly as before.

During the period that the *Compositio* was issued, England witnessed the growth of Parliament and the beginning of statute-making, but their immediate effect on weights and measures legislation was not noticeable

---

9. Even though the inch is defined repeatedly throughout medieval English weights and measures legislation as the length of 3 barleycorns, the actual standard in fact was a particular rod of metal, usually a yard-bar, on which inches were marked.

until the very end of Edward I's reign. Earlier, in 1296, Edward had decreed that the London quarter mentioned in Magna Carta be fixed at a capacity of 8 striked bushels and in 1303 his government issued the *Tractatus*.[10] His decree of 1296 is unique because hitherto the London quarter had never been defined in terms of its aliquot parts. The provisions of the *Tractatus* dealing with the tower pound and with measures of length and area were not new, as evidence from the *Compositio* and Henry's assize proves. But the document described two capacity measures for the first time. It declared that 8 tower pounds of wheat made the gallon of wine and that 8 gallons of wine made the London bushel. The wording of this particular clause has produced a great deal of confusion. The first statement actually means that a cask filled to the brim with 8 pounds of wheat will have a total cubic capacity sufficient for a wine gallon. It does not mean that the wine, with or without the cask, will be equivalent in weight to 8 pounds of wheat. The last half of the clause means that 8 casks, each filled to the brim with 8 pounds of wheat, will produce collectively a cubic capacity sufficient for a bushel. Of course, in the latter case the total weight of wheat in the bushel will be 64 tower pounds. English legislation was still several centuries away from defining capacity measures in terms of cubic inches or even in terms of diameters and circumferences.

There were very few innovations in the legislation of Edward II (1307-27). Parliament in 1324 legalized by statute the lengths of the linear measures found in the *Compositio* and during the course of the next year issued the Ordinance for Bakers, Brewers, and Other Victuallers and the Ordinance for Measures. The first document forbade heaped or shallow capacity measures for all goods except oats, malt, and meal, and emphasized that all standard measures had to be stamped with the iron seal of the king. The removal of heaped and shallow measures of oats, malt, and meal from the general prohibition is the earliest legislative example of protecting certain local privileges and vested interests at the

---

10. All dry capacity measures were either heaped (*comble, coumble, cumulatus*), striked (*ras, rasa, rasyd, sine cumulo* or *cumulata, stricke, stryke*), or shallow (*cantel, cantell, grains sur bords*). The heaped measure contained an amount of grain extending above its rim. The actual amount in excess of a level measure depended on the proportions of the vessel. If two vessels contained an equal capacity, yet one was shallow with a wide diameter and the other was deep with a narrow diameter, the former, if both were heaped, would contain the greater amount. A vessel in which the contents did not extend above the rim was a striked measure. A measure was striked by passing a straight piece of wood called a streek, strike, or strickle over and along its rim in order to remove any excess grain. The striked measure was thus a level measure. A shallow measure was one in which the contents did not reach the rim. The vessel either was purposely filled this way or the merchant or seller compressed its contents below the rim.

expense of standardization. Local merchants relished such grants. For example, they preferred doing business at markets employing excessive measures, for they could purchase their supplies and pay taxes in larger units and then sell them in smaller ones. In other words, they made a greater profit in these markets than in those employing the king's standards. The insistence on stamping all state and local standards with the king's seal originated, as far as I can determine, with William I. But even in this clause indecisive language crippled well-intentioned efforts. It did not mention the officers charged with the custody of the standards nor the places where they were to be kept. It merely threatened offenders with a fine of £100, an extremely large sum when compared with fines in similar legislation up until relatively recent times. The Ordinance for Measures was simply a repetition of those sections in the *Tractatus* dealing with the tower pound and with capacity measures. Nothing new was added.

The new standard of linear measurement outlined in the *Compositio* became the basis after the thirteenth century on which cloth dimensions were defined. In the earliest application of this yard to the measurement of textiles a statute of Edward III (1327–77) issued in 1328 fixed the length of ray, or striped, cloth at 28 yards and its width at 5 quarters, while colored cloths were to measure 26 yards by 6½ quarters. A quarter in the cloth trade was a measure of ¼ yard, or 9 inches. Statutes in 1351, 1371, and 1373 reconfirmed these provisions except that the last enactment altered the width of colored cloth to 6 quarters.

Aside from these cloth regulations, weights and measures law during the reign of Edward III was concerned principally with the elimination of illegal metrological practices. For instance, an act in 1340 prohibited buying products in large local measures and reselling them in smaller statutory sizes. This was directed specifically at those merchants who reaped additional profits by such practices and who, in the process, swindled the government out of a sizable share of its tax revenues. Too, it could be argued that this document was an attempt to rescind the privilege granted in the Ordinance for Bakers, Brewers, and Other Victuallers, which authorized the sale of oats, malt, and meal in heaped capacity measures. But an enactment issued eleven years later would seem to deny this. In the Statute of Purveyors of 1351 Parliament demanded that all bushels, half-bushels, pecks, gallons, pottles, and quarts be made according to the king's standards; prohibited weighing of goods in markets by the auncel scale;[11] and insisted that all measures of corn be striked except

---

11. See my *Dictionary* (Madison, 1968), sv *auncel,* and also "Seventh Annual Report of the Warden of the Standards on the Proceedings and Business of the Standard Weights and Measures Department of the Board of Trade for 1872-73," in *Parliamentary Papers,*

in those cases where local lords had used other types of measures on their estates that were at variance with Crown standards. Besides failing to define the specifications for units and standards, this was the second time within 25 years that certain groups were granted the right to ignore statutory provisions. Just how many manorial lords claimed this particular privilege is impossible to determine. It is reasonable to assume, however, that most of them could argue that they had always employed measures that were different from Crown standards. Parliament was yielding to local custom and this represented a direct contradiction of the work it was trying to accomplish.[12]

All of the Edwardian statutes after 1351 were basically amendments to previous legislation. The weights and measures provisions in the Statute of Purveyors were repeated in 1353 in chapter 10 of the Statute of the Staple. This document complained that the injunction of 1340 prohibiting merchants from buying in large local measures and reselling in smaller statutory ones was being totally ignored. Illegal use of the balance was forbidden and a heavy fine and prison sentence awaited anyone found guilty of fraudulent weighing practices.[13] In 1357 the Statute of West-

---

Great Britain, *Reports from Commissioners*, 38 (London, 1873): 55; *Select Cases concerning the Law Merchant: A.D. 1239-1633*, ed. Hubert Hall, Selden Society Publication, 46 (London, 1930): 76; Frederick T. Dinsdale, *A Glossary of Provincial Words Used in Teesdale in the County of Durham* (London, 1849), p. 137; *The Charters of Endowment, Inventories, and Account Rolls, of the Priory of Finchale, in the County of Durham*, Surtees Society Publication, 6 (London, 1837): ccccxlix, ccgclii; *The Fabric Rolls of York Minster*, Surtees Society Publication, 35 (Durham, 1859): 355; Bruno Kisch, *Scales and Weights: A Historical Outline* (New Haven, 1965), pp. 58-60; and *Munimenta gildhallæ Londoniensis: Liber Albus, Liber Custumarum et Liber Horn*, ed. Henry Thomas Riley (London, 1859-62), p. 586.

12. It is interesting to note that after the passage of this statute, many governmental directives to local sheriffs were similar to the following one, found in the *Calendar of the Close Rolls: Edward III: 1354-1360* (London, 1908), p. 226: "To the sheriff of Devon. Order to cause proclamation to be made that all who have measures shall bring them to the sheriff and have them made to agree with the standard and that no one, upon pain of forfeiture, shall use any measures except streeked measures agreeing with the standard, as a statute [the Statute of Purveyors] passed by common counsel of the realm of England contains that all measure[s] . . . shall be made to agree with the standard throughout England . . . and each measure of corn shall be streeked, saving the rents and ferms of lords which shall be measured by the same measure as heretofore."

13. The balance was first mentioned in the Statute of Purveyors as the only legal scale for trade. It consisted of a beam or lever supported by a fulcrum at the midpoint to form two equal arms. A pan was suspended from each arm, one to hold an object of known weight and the other to hold the goods to be weighed. The weight of the goods was determined either when equilibrium between the two pans was achieved or when the weight of the goods was registered by the deflection of a pointer fastened to the beam and provided with a weight chart before which it moved back and forth. Unlike the auncel scale, there was no lifting of the instrument whatsoever. The fraudulent weighing practices mentioned in the statute of

minster I stated that since illegal weights, measures, and scales continued to be employed throughout England, it was mandatory that merchants and producers bring their standards and weighing apparatus to their country sheriffs for testing. Refusal to comply with this order resulted in punishment by fine at the king's will. The units mentioned, but never defined, were the sack, half-sack, quarter-sack, pound, half-pound, and quarter-pound. Even though the units were not defined, there is substantial evidence in contemporary sources that the weights mentioned were avoirdupois. Established primarily for wool weighing sometime during the early years of Edward's reign, the standard for the avoirdupois pound until the time of Elizabeth I contained 6992 grains, or 16 ounces of 437 grains. This ounce was the same as that found in the medieval Roman and Florentine weight systems. There is no better example than this of the effect of international trade on the expanding British system of weights and measures.

The last of these amendatory enactments was passed in 1360. Once again Parliament forbade continued use of the auncel scale.[14] But the

---

1353 resulted when the balance was tipped slightly to one side or the other or when the merchant's hand or foot came into contact with the instrument during the process of weighing. By the terms of the statute, offenders had to forfeit the value of the goods to the king and four times their value to the customer in addition to serving a prison sentence of one year. While practically every market after 1353 purchased balances, in the large ports there was a "king's beam" whose use was compulsory for all goods sold in bulk, whether imports or exports. It would be more correct to say that there were two "king's beams," for the "great beam" for weighing wool and other heavy goods by the hundredweight was distinct from the "small beam" used for weighing by the pound such costly luxury items as silk. These beams or balances were called "trones" or "tronas" and the persons entrusted with their operation were "tronators" or "tronours." Although their use was compulsory there were fees to be paid and the revenue derived from them was considerable. Like weights and measures standards, the beams were distributed from the Exchequer. See also L. F. Salzman, *English Trade in the Middle Ages* (Oxford, 1931), pp. 59-61; "Seventh Annual Report of the Warden of the Standards," p. 45; *Munimenta gildhallæ Londoniensis,* pp. 588-89; Hubert Hall and Frieda J. Nicholas, eds., *Select Tracts and Table Books Relating to English Weights and Measures (1100-1742),* Camden Third Series, 41 (London, 1929): 42-43; *Memorials of London and London Life in the XIIIth, XIVth and XVth Centuries,* ed. Henry Thomas Riley (London, 1868), pp. 26, 72, 74; and *The Coventry Leet Book,* ed. Mary Dormer Harris (London, 1913), p. 396.

14. Part of the blame for the continuing use of the auncel scale can be placed directly on the government, for London acted very slowly in supplying the counties with the proper weighing devices. The following entry in the *Calendar of the Close Rolls: Edward III: 1360-1364* (London, 1909), p. 183, is an example of this: "To the justices appointed in Buckinghamshire to inquire concerning weights and measures. Order to suffer John de Claydon, Henry Glover of Stretford . . . merchants to weigh their wool and other merchandise by the weight called auncel, and to use that weight until Christmas next, that in the meantime the king may cause balances and weights agreeing with the standard of the exchequer to be prepared and sent to the sheriff according to the statute, not troubling the

document reconfirmed the privilege extended to manorial lords in 1351 of using measures that did not conform to Crown standards. It is indeed strange that such a privilege, which had no time limitations, would have to be reconfirmed, but the government, probably harassed by those persons or groups who were not included in the grant, wanted to reaffirm its initial position that only certain manorial lords were entitled to the grant's benefits. Since merchants traditionally had used measures that did not conform to Crown specifications, they, for one, wanted to be included. Parliament seemed to be saying to them that feudal custom was more important than commercial custom.

Edward III died in 1377 and was succeeded by the last monarch of the Angevin-Plantagenet line, Richard II. Even though he was overburdened with prosecution of the war with France that had begun in the early years of his father's reign, Richard (1377-99), together with his parliaments, enacted four major weights and measures laws. In the first, issued in 1380 at Northampton, Parliament acknowledged that it had never established regulations for wine measures imported into England, Wales, and Ireland. Without even mentioning the names of the measures, Parliament ordered the king's gaugers, or port measurers, to check carefully all incoming wine casks and to punish according to the provisions of past enactments all importers who refused to surrender their casks for inspection. The act also threatened gaugers with punishment according to past provisions if they failed to abide by the new regulations. The casks referred to were probably tuns and pipes—the most common wine measures in British international trade—but aside from brief mention of them in the Edwardian statutes of 1353 and 1357, they had never been defined by law. The 1353 enactment ruled that if tuns and pipes were found to be deficient in size, the monetary value of the deficiency was to be deducted in the final payment to the importer; according to the 1357 statute, any person selling a tun or pipe of wine not duly gauged was to forfeit the wine or its monetary value to the king. If gaugers had never received official instructions concerning the acceptable dimensions of tuns and pipes, then decisions as to whether they were legal rested on their personal judgments.

The most important legislative act of Richard's reign was passed in 1389. There were three principal provisions. First, all illegal weights and measures in England were to be burned, the process to be supervised by the clerk of the market for the king's household. Parliament was trying

---

said merchants for reasonable use of the said weight hitherto, provided that they be severely punished if they have committed any fraud or deceit; upon the petition of the said merchants praying the king to grant that, whereas such balances and weights are not yet sent to that county . . . they may in buying and selling use the weight called auncel, as they were heretofore wont to do, until the same be sent."

desperately to prevent merchants and producers from hastily repairing illegal weights and measures to pass government inspections and then afterward counterfeiting them to suit local preferences. A larger than statutory measure, for instance, could be installed with a false bottom to render dimensions in agreement with Crown specifications; after the inspection this could be removed. The act also fixed the stone of wool at 14 pounds and the sack of wool at 26 stone, or 364 pounds. This rating reflects the importance of England's export trade because the sack was adapted both to the 500-librae standard (360 BI pounds) used by the Florentine merchants and to the skippund (ship pound) of the Baltic countries and of Scotland, which consisted of 20 lispund, containing 16 Norse troy pounds each, or 320 Norse troy pounds in all (approximately 352 BI pounds). Last, the document permitted Lancaster continued use of its excessively large measures. The official explanation was that this county had always maintained measures that were considerably larger than those found anywhere else in the realm and that to replace them with statutory measures would be needlessly expensive. What probably motivated Parliament in extending the privilege was the powerful position of the Lancastrians in government affairs and the pressure that they applied on Richard's ministers. Lobbying for vested interests has existed in all ages.

The last of Richard's statutes appeared in 1391 and 1393. The first prohibited the London fatt and reemphasized that the legal quarter was to contain no more than 8 striked bushels. The fatt had the same physical dimensions as the quarter but it was a heaped rather than a striked measure. Its 8 heaped bushels were equivalent to 9 standard striked bushels. Hence, it was a heaped quarter.[15] The 1393 statute was a repudiation of all earlier cloth statutes dating back to Magna Carta, for the act provided that merchants could sell cloth of any length and width as long as they paid the necessary alnage fees and excise duties. This privilege would not survive, but Parliament, at least in 1393, admitted that the goal of regulating cloth measurement was unattainable.

Lancastrian Legislation

The act of 1393 provided the impetus for a confusing series of cloth statutes during the reign of Henry IV (1399-1413), the first of the Lancastrian monarchs. In 1405 Parliament annulled the provisions of the statute of 1393 and renewed Edward's regulations of 1328 save that all rays and colored cloths were now to be uniformly 28 yards in length.

---

15. See Salzman, *English Trade*, p. 46, and *Rotuli parliamentorum ut et petitiones, et placita in parliamento tempore Ricardi R. II*, ed. Rev. John Strachey et al., 3 (London, 1832): 291.

Anyone who disobeyed these new regulations forfeited his cloth to the alnager making the inspection. The alnager, in turn, was instructed to send the cloth to the king for his personal use. Two years later Parliament repealed this enactment and declared that standard sizes for cloth were no longer necessary. Clothmakers were again free to choose any dimensions they thought appropriate. The statute explained that Henry considered the former restrictions to be prejudicial and he reportedly apologized for any inconvenience suffered by clothmakers and merchants. This act survived for two years. In 1409 Parliament repealed it and reconfirmed the 1405 enactment. A statute of 1411 then repeated the regulations of 1405 and 1409. Standardization was weakened seriously by continuous Parliamentary indecision. Enactments obviously had to be repealed if they were proven ineffective or unfair, but when they were repealed and then reinstated almost systematically, Parliament created considerable confusion for merchants and their customers and made the task of nationwide enforcement of current statutory provisions much more difficult to achieve.

Two statutes of Henry V (1413-22) are noteworthy. In 1414 the troy pound was mentioned for the first time, in an act dealing with the price and composition of English goldsmiths' silver-gilt work. The various units in the troy system were not defined even though they had been commonly used in England since the late fourteenth century. The system is also mentioned in an act of Henry VI's in 1423 regulating the price of bar and plate, but even here there is no description provided. In both statutes it is accepted only as a legal basis for weighing and never as a replacement for the tower system. Contemporary documents indicate, however, that the troy pound contained 5760 grains (373.242 grams), and consisted of 240 pennyweights of 24 grains each, or of 12 ounces of 480 grains each. One of the reasons for its quick acceptance by merchants was that exchanges between it and the tower system and between both of these and the northern German system were relatively simple. For example, 15 troy ounces of 480 grains equaled 16 tower ounces of 450 grains, both being equal to 7200 grains, the weight of the North German pound that had been introduced into England by the Hanseatic League. The name for the new system probably came from the French marc of Troyes, but it is certain that the English troy pound did not come from Troyes, France. The Troyes marc had an ounce equal to 472.1 BI grains. There was a family of pounds usually known as troy in the northern trade of the period whose ounces varied from 483 to 472 grains.[16] The English troy pound took its

16. Specifically, the Swedish mark-weight pund had an ounce of 483.3 BI grains; the Danish solvpund ounce, 481.5 BI grains; the Scots tron pound ounce, 481.1 BI grains; the Bremen pound ounce, 480.8 BI grains; the Norwegian skaalpund ounce, 477.4 BI grains; the

name, like the Scots and Dutch pounds, from the Troyes marc, but took its standard from some pound of full weight, probably the Bremen pound whose ounce weighed 480.8 BI grains.

The chalder and the keel—the principal coal measures—are mentioned initially in a statute of 1421, but the act did not describe what their dimensions or their relationships were. It can be seen from contemporary fifteenth-century documents that the chalder's contents weighed one ton and that 20 of these chalders made the keel-load or barge-load everywhere except in Newcastle-upon-Tyne, where the chalder totaled 42 hundredweight (5936 pounds) and equaled ⅛ keel. But it became customary to build larger-sized keels so that additional chalders could be loaded. Shippers thus minimized customs rates, which were assessed usually on each keel rather than on the individual chalder. Hence the statute established a customs duty of 2d. on each and every chalder, and it was assigned while the chalders were being transported on keels from the docks to merchant ships. The document also declared that the keels were to be examined by commissioners appointed solely for this purpose and the vessel's total load was to be registered. Failure to comply with these instructions resulted in forfeiture of the entire contents. Eventually shippers evaded these regulations by enlarging the size of the chalder so that by the late seventeenth century its weight had increased considerably: 240 pounds more in the regular chalder, and 1232 pounds more in the one from Newcastle.

The last important pre-Tudor statutes were enacted during the reign of Henry VI (1421-61). In 1423 Parliament promulgated unit standards for 6 wine and fish measures. All of them were defined in terms of their gallon capacities but the document did not give any definition of the gallon itself. Hitherto the only gallon mentioned was the one for wine in Edward I's *Tractatus,* where it was equivalent to the cubic capacity of a cask large enough to hold 8 tower pounds of wheat. Obviously, Parliament was referring to some physical standard housed at one of the state depositories in London but since none of the pre-Tudor standards for capacity measurement has survived one can only assume that the units mentioned in Henry VI's statute were approximately equivalent to those issued by Henry VII and Elizabeth I. In the following list the metric equivalents of the 1423 standards are based upon these Tudor standards.

herring and eel barrel	=	30 gallons fully packed (ca. 1.14 hektoliters)
salmon butt	=	84 gallons fully packed (ca. 3.18 hektoliters)
wine hogshead	=	63 gallons (ca. 2.38 hektoliters)

Amsterdam pound ounce, 476.6 BI grains; the Scots trois pound ounce, 475.5 BI grains; the Dutch troy pound ounce, 474.7 BI grains; and the French troy pound (Troyes marc) ounce, 472.1 BI grains.

wine tertian	=	84 gallons (ca. 3.18 hektoliters)
wine pipe	=	126 gallons (ca. 4.77 hektoliters)
wine tun	=	252 gallons (ca. 9.54 hektoliters)

Six years later a statute confirmed the document of 1351 that had abolished the auncel scale. This third statutory condemnation (the second was in 1360) of auncel weighing carried with it detailed provisions for distributing standard balances and weights throughout every city, borough, and town in England. All citizens could use the balances free of charge, but foreigners had to pay fees based upon the total weight of each weighing. The elimination of the auncel was not achieved with this enactment. So rooted was the use of the unequal-arm scale among the common people that it was still found in England in the eighteenth century.[17] Even the threat of excommunication from the church for anyone found guilty of using the auncel, promulgated by the Archbishop of Canterbury in the fifteenth century, did little to curb its popular appeal.

The Emergence of Standards

The underlying purpose of these many medieval enactments was the creation of a coordinated, uniform system of weights and measures. From the twelfth century to the end of the fifteenth, significant strides in this direction were made. That ultimate success was not achieved was due to more than the inherent deficiencies in the laws. A myriad of problems concerning physical standards and the officials responsible for weights and measures contributed to the overall weights and measures dilemma which the Tudor governments would inherit after 1485.

Constructed throughout the Middle Ages in London, Edinburgh, Aberdeen, and Perth, physical standards were sent by parliamentary or royal directives to all shires for distribution among the principal urban centers.[18] To check against frauds, local authorities policed their safe-

---

17. See *Dictionarium rusticum, urbanicum et botanicum: or, A Dictionary of Husbandry, Gardening, Trade, Commerce, and All Sorts of Country-Affairs* (London, 1717), sv *auncel.*

18. *Munimenta gildhallæ Londoniensis*, p. 382: "Clericus Marescalciæ Domini Regis, detulit apud Gildaulam mensuræ Domini Regis, ad assiandum per standardum civitatis" ["The clerk of the market of the lord king takes the measures of the lord king from the Guildhall in order to assay the standards of the city"]. *Rotuli parliamentorum et petitiones, et placita in parliamento tempore Edwardi R. III,* ed. Rev. John Strachey et al., 2 (London, 1832): 260: "Qe plese a *notre* dit Seign' le Roi d'ordiner, q*ue* Busseux d'areine soient faites en la Tour de Loundres, pur chescun Countee d'Engleterre . . . apres quel touz les Busseux de cel Countee serront faitz & mesurez" ["That it pleases our said lord, the king, to order that brass bushels be made in the Tower of London for each county of England . . . after which all bushels of each county will be made and measured (sized)"]. *Coventry Leet Book*, p. 133: "The said maiour sent to London for to haue the weightes acordyng to the weightes of thexechecour." *Calendar of the Fine Rolls: Edward II: 1319–27* (London, 1912), p. 315:

keeping, inspected them periodically, and stamped them according to statutory specifications. In many instances they were attached to the walls of local guildhalls or other municipal buildings. Their use in the larger ports was especially important because merchants and artisans gathered in these commercial depots to conduct their business transactions. Having convenient access to copies of Crown standards also helped the local guild or market control illegal practices.

But special emphasis must be placed on the fact that these standards were not originals. They were copies. Sometimes they were duplicated with accuracy and precision, yet more often than not they varied appreciably from the originals. In addition, local copies frequently were made from other copies, and if local craftsmen were less talented than those employed by the monarch or Parliament, the disparity between state and local weights and measures grew more acute.

Further, local standards tended to deteriorate rather quickly since they were in constant use. Wear and tear, however, was not the only cause of their inaccuracy. Wooden standards built by the turners naturally decayed or became worm-eaten, and those made from lead, iron, or brass oxidized.[19] Also, since they were not kept under controlled atmospheric temperatures or in hermetically sealed compartments, they were continually expanding and contracting depending on seasonal and climatic changes. A brass yard-bar, for example, would be measurably shorter in winter than in the summer. Modern platinum and gold standards are far more accurate and reliable than the former iron, brass, lead, and wooden models, and they are affected much less by the perils of occasional cleaning and repeated handling or by the ever-present threat of air pollution.[20]

"Afterwards the king caused to be made certain measures . . . by the said standard, and to be sent to the principal town of each county . . . that certain measures in those counties might be made and proved by the measures so approved." *The Maire of Bristowe Is Kalendar by Robert Ricart, Town Clerk of Bristol 18 Edward IV,* ed. Lucy Toulmin Smith, Camden Society Publication, New Series, 5 (Westminster, 1872): 84: "Item, that all maner of colyers that bryngeth colys to towne for to sille, smale or grete, that they bryng their sakkes of juste mesure, according to the standard, for the which the maire is vsed this quarter to commaunde the standard mesure[s] to be sett in diuers places of this Toune, as at the High Crosse, the Brigge Corner, and Stallage Crosse, so that every sak be tryed and provid to be and holde a carnok." The Wardrobe in the Tower of London and the Standards Department in the Exchequer preserved regalia in addition to weights and measures standards.

19. See *Memorials of London,* p. 78; "Seventh Annual Report of the Warden of the Standards," p. 49; Kisch, *Scales and Weights,* p. 84; Salzman, *English Trade,* pp. 56-57; and Ephraim Chambers, *Cyclopædia: or, An Universal Dictionary of Arts and Sciences* (London, 1728), 2:360.

20. *Calendar of the Close Rolls: Richard II: 1381-1385* (London, 1920), p. 365: "To the mayor and constable of the staple of St. Botolph, the collectors of customs and the

Medieval Times

Local and state standards of pre-Tudor vintage are as rare as those surviving from Anglo-Saxon times. In the Westgate Museum in London there is a 91-pound wool weight that dates from the early years of Edward III's reign. Constructed, conceivably, soon after the statute of 1328 had established governmental specifications for the stone and sack of wool, this unusual weight represented exactly ¼ of a sack of wool, which totaled 364 pounds. It is the only one of its kind ever found in England. Since weights larger than one hundredweight would be extremely difficult and cumbersome to handle, it was customary in weighing large articles to add a multiple of smaller weights to the steelyard, or balance, until equilibrium or other desired results were obtained. Consequently, four of the 91-pound weights would be used in any given weighing process to determine whether the sack conformed to the legal 364-pound capacity. Similar techniques were used in ascertaining the correct statutory weights of chalders, keels, tons, fothers, weys, and other inordinately large weights and measures.

In 1357 an Edwardian statute commanded that certain balances and weights for wool built to Exchequer specifications be sent to all English sheriffs. This must have been the date when Winchester received from the Exchequer its set of avoirdupois weights in the series of 56 pounds, 28 pounds, 14 pounds, and 7 pounds. All of them bear the Royal Arms quartering of Old France, only used between 1340 and 1405.[21]

During this same period the wool trade flourished at the Great Fair of Winchester—the center for the buying and selling of England's most

---

controller in that port for the time being: Order, upon petition of merchants of that town and others flocking to the staple there, if their complaint is true, from time to time when need be to amend all weights appointed of old time for weighing of wool therein which by frequent use are worn so light that they agree not with the standard as by the merchants it is found, that the king be not defrauded nor the merchants." Nothing was accomplished since the same plea was made several times over the next decade. *The York Mercers and Merchant Adventurers: 1356–1917,* Surtees Society Publication, 129 (Durham, 1918): 309-10: "The wardens acquaint the court that there have been complaints against our standard weights and that they are worn light and wrong, upon which the court requests Mr. Governor to write to London for a new sett according to the standard, which his worship promised to doe." Adrien Fauve, *Les Origines du système métrique* (Paris, 1931), p. 16: "[Ils] était exposé[s] aux chocs, aux injures de l'air, à la rouille, au contact de toutes les mesures qui y sont présentées et à la malignité même de tout malententionné" ["(They) were exposed to shocks, to air damage, to rust, to contact with all the measures by which they were (assayed) and even to the malignity of the most ill-intentioned"].

21. Note that the doubling of these weights, beginning with the basic weight of 7 pounds, produced a quarter-hundredweight of 28 pounds, a half-hundredweight of 56 pounds, and a hundredweight of 112 pounds. This helped to oust the decimal series of 25-50-100, which was from that time on reserved almost exclusively for heavy metals.

essential commercial product for export. Held every September on St. Giles's Hill, the fair attracted, from all over the country, merchants who came to sell wool in addition to cloth, pottery, wine, and spices. Here, also, were found the iron weights of Edward III employed in settling the countless disputes arising between buyers and sellers. These weights were made of lead and had a loop at the top through which a leather thong passed. Thus a pair of them could be slung conveniently across the back of a horse on which the troner (or tronator) rode as he verified weights and collected dues. Some of the original leaden seals used for authenticating sacks of wool and cloth after weighing can be seen at the Westgate Museum.[22]

Little else is available from this era. There are a few documentary references to physical standards but no extant models for the period between the reigns of Richard II and Henry VII. In 1393 York reportedly had the following standards:[23]

Designation	Material
grain bushel	eris [copper or bronze]
grain half-bushel	wood
grain peck	wood
ale gallon	wood
ale pottle	wood
ale ⅓ gallon	wood
ale quart	wood
wine gallon	eris [copper or bronze]
wine pottle	eris [copper or bronze]
wine quart	eris [copper or bronze]

22. For discussions and historical citations concerning seals and their innumerable forms and applications, see A. E. Berriman, *Historical Metrology* (London, 1953), p. 168; H. S. Chaney, *Our Weights and Measures, A Practical Treatise on the Standard Weights and Measures in Use in the British Empire with Some Account of the Metric System* (London, 1897), p. 51; *York Memorandum Book: Part I, (1376-1419)* Surtees Society Publication, 120 (Durham, 1912): xxiii, lxvi-lxvii, 15-16, and *Part II,* 125 (Durham, 1915): 90-91; Kisch, *Scales and Weights,* pp. 163-64; *The Whole Volume of Statutes at Large, which at Anie Time Heeretofore Have Beene Extant in Print* (London, 1587), pp. 77, 429, 430, 433-34; *Munimenta gildhallæ Londoniensis,* p. 285; Hall and Nicholas, eds., *Select Tracts,* p. 47; *The Statutes of Ireland, Beginning the Third Yere of K. Edward the Second, and Continuing untill the End of the Parliament, Begunne in the Eleventh Yeare of the Reign of Our Most Gratious Soveraigne Lord King James* (Dublin, 1621), p. 18; *Memorials of London,* pp. 255-56; Chambers, *Cyclopædia,* 2 (1728): 358-60; and *A Collection in English of the Statutes Now in Force, Continued from the Beginning of Magna Charta, Made in the 9. Yere of the Raigne of King H. 3. until the End of the Parliament Holden in the 7. Yere of the Raigne of Our Soveraigne Lord King James* (London, 1615), p. 463.

23. *York Memorandum Book, Part II,* p. 13.

A list of standard measures kept at Beverly in 1423 reveals:[24]

Designation	Material
pottle	pewter
quart	pewter
pint	pewter
gill	pewter
panniers	wood
"hopir"	wood
modius	wood
"firthindal"	wood
gallon	wood
pottle	wood
⅓ gallon	wood
quart	wood

Finally, the mayor of Coventry ordered in 1434 that all strikes of the city be made according to the standard of King John's time, "the which standard is . . . in the maiouris almery in the geyl-hall off Couentre."[25] In a later section the mayor was instructed to order strikes of latten and brass, but no mention was made as to where they were coming from or who their makers were. Since the strike was a local and not a statutory measure, it is unlikely that London was the supplier.

Officials and Enforcement

By themselves, statutory injunctions and physical standards cannot guarantee uniformity in weights and measures. A well-trained and dedicated corps of officials is needed to enforce the laws and to inspect and verify local and national standards. These separate, but interrelated, tasks of inspection, verification, and enforcement constitute the axis around which the entire weights and measures program revolves. If officials perform their assigned duties well, the goal of a standardized metrology will be realized. If they are inadequately trained, unknowledgeable concerning the important aspects of their work, or even corrupt, laws will go unobserved and physical standards will represent merely the scientific expertise of their inventors.

Unfortunately, the above adverse situations prevailed all too often during the Middle Ages. The sources describe large numbers of people, both urban and rural, utilizing the weights and measures function for self-aggrandizing economic or political considerations. During the fourteenth and fifteenth centuries, for example, there is constant agitation for the removal of the principal abuses. The dissatisfaction is voiced by

24. L. F. Salzman, *English Industries of the Middle Ages* (Oxford, 1923), p. 288.
25. *Coventry Leet Book*, p. 151.

Commons in their petitions to king and council, and it is met by emergency legislation and appointments of justices *ad hoc,* followed by the revoking of the commissions and a frantic appeal for the selection of justices of the peace. Besides the various justices, there were presentments made before local authorities, mayors, bailiffs, and stewards in staple courts, tolzey courts, courts leet, and courts baron. There were appointments of guildsmen, manorial lords, coroners, clerks of the market, abbots, priors, and archbishops. The list could be extended much further, but the point is that with too many individuals and groups involved, the possible sources of fraud and corruption were augmented significantly. Often their duties were poorly defined or not defined at all, and they continually encroached on each other's territories and jurisdictions. In almost every case one's remuneration depended on the number and amount of fines levied, and this increased the likelihood of overzealousness on the part of the administering official. It was also customary to obtain special privileges in local legislation enabling proprietors of fairs and markets to regulate their own weights and measures. In this way the statutory system of inspection was weakened and still more names were added to the already overcrowded rolls of weights and measures personnel. Under such conditions and with rivalries so prevalent among those chosen by the Crown and Parliament to carry out standardization, it is not surprising that the road to metrological reform was a long and arduous one. Not until the nineteenth century was this most important area of metrological control overhauled and restructured to produce maximum efficiency with a minimum of manpower and resources.

Urban Residents

Before the reign of Richard I there is no record indicating that state and local governments delegated to anyone the functions of inspecting and verifying weights and measures or of enforcing decrees pertaining to their control. None of the extant documentary materials indicates whether such functions fell within the normal duties of bureaucratic officials at London, or earlier at Winchester; whether shire, hundred, or other courts customarily handled such matters; or whether Richard was the first English monarch to inaugurate this particular area of weights and measures regulation. If either the first or second supposition was true, then the state perhaps did not feel it necessary to record duties that were associated traditionally with legislative, financial, and judicial officials. If the third was true, then Richard's contribution is truly outstanding. On the one hand, he may have recorded merely what custom had dictated for centuries, but on the other, he may have begun a tradition whose implications are monumental in the history of British metrology. Whichever the case, the administration of weights and measures may be said to have been formally set into motion with his Assize of Measures in 1197.

Richard's assize represented the most modest of beginnings. Although it allocated responsibility for the assize of the king's standards to certain people in every city and borough of the realm, it leaves too many important questions unanswered. Who were those selected? How many of them were there altogether or, at least, what was the number assigned to each urban center? From what municipal or social groups were they drawn and what constituted their length of service—were they appointed for life or were they subject to periodic reassessment? What were their responsibilities? Were they merely examiners or could they verify local weights and measures, authenticating those that conformed to Crown specifications by stamping them with Exchequer or other seals? Did they perform their duties individually or collectively? In either case did they correlate their functions with those of elected or appointed urban magistrates? Finally, did their responsibilities extend to enforcement of penal codes or was this area of jurisdiction reserved solely for the king's justices, itinerant judges, or local judicial personnel? Unfortunately, the assize provides no clues and other contemporary documents never mention such matters.

Some of the mystery surrounding certain aspects of this assize was removed seven decades later. In the Assize of Bread and Ale of Henry III, provision was made for six "lawful" men to gather the weights and measures of their town and to test them for possible variances with the Crown's standards. It was mandatory that each local weight or measure be inscribed in legible script with its owner's name. After the examination the commissioners were to turn over to a panel of twelve other men all local standards that did not pass the initial inspection. This jury, acting on the king's behalf, would determine ultimately whether the weights and measures presented to them were of tolerable variances or whether they were illegal for commercial transactions. If the latter was decided the jury had the right of punishing the guilty parties.

Even though several of the problems pertaining to Richard's assize were eliminated clearly and decisively, most were not, and to further complicate the issue, the assize of 1266 created some new problems because of its incomplete or imprecise language and because of conflicting evidence found in contemporary documents. First, "lawful" men could have at least three specific and unrelated meanings. It could imply men with respectable standing whose good reputations were known to all in a community. Such men would have been law-abiding and considered dependable, dedicated, and honest. At the same time, however, the term might designate those urban residents selected legally either by the government in London or by local politicians; in other words, only those who were properly appointed or elected could serve. It could also relate to those city dwellers learned in the law, indirectly implying the type of personnel frequently empaneled to serve on sworn inquests and grand juries. Of these three possibilities, the second seems the most likely. Such men, it can be assumed, would have had the attributes ascribed to those citizens in the first interpretation since such traits have always been basic qualities demanded of public servants. Also, they were probably knowledgeable concerning the law since their function required familiarity with the specifics of the assize and entailed some understanding of local laws, customs, or traditions. No more conclusive a case than this can be made.

Second, it is never stated whether the commissioners' functions ceased once they relinquished their weights and measures to the jury. Did they begin a new round of examinations or terminate their investigations until the next year or the next scheduled testing date? Did they remain available to the jury as witnesses against the defendants or as suppliers of testimony needed by the jury in formulating its final decisions? Which of the two groups was responsible for insuring appearances of the accused at the trials and how were such people summoned—by writ or by other methods?

Finally, how could the jury punish? Was there a regular scale of fines promulgated by the central government, by local magistrates, or were such matters left to the jury's discretion or to the dictates of local custom? Were such financial settlements the exclusive property of the Crown or were they divided in some fashion? If fines were not levied, could the jury order corporal punishment? If so, what kinds were allowed and who administered the actual police action?

Other documents, contemporaneous with Henry's assize, do not add significantly to one's understanding of the questions raised above but they do indicate, unmistakably, that the enactment was not rigidly enforced. In every entry in the Patent Rolls dealing with appointments of weights and measures personnel for the period between 1266 and 1272, no more than two were given metrological functions in any designated region. Furthermore, such men were responsible for a geographic and demographic area much larger than a city or borough. They were always assigned a minimum of one shire, more frequently two. Once appointed, the commissioners were to "view, prove and measure" bushels, gallons, ells, and weights. In one entry they could "amerce tre[s]passes of the said measures within liberties and without"; in a second they were to work in close collaboration with bailiffs appointed by the king; in yet a third they were instructed to "amerce tre[s]passers . . . with mandate to all bailiffs, reeves, and commonalities of boroughs and market towns . . . to be intendant to them."[26] These examples hardly coincide with the specifics of the assize and they quite possibly attest to the extremely precarious position of royal demands and conciliar enactments in the face of local preferences and customary practices. Furthermore, one cannot determine from any late-thirteenth-century materials whether the assize was ever officially promulgated; whether the government altered its administrative regulations immediately following its issuance; whether London was forced to amend it substantially because of cost or time factors; or whether local resentment precluded its application.

The administrative practices described in the Patent Rolls may have been at variance with Henry's assize, but they became standard procedure by the fourteenth year of Edward III's reign.[27] In a statute of 1340, Parliament declared that after standards of the bushel, gallon, and sundry weights had been distributed to shires that hitherto had been denied them, two "good and sufficient persons" were to be assigned in every shire to "survey" all weights and measures. Specifically, the new commissioners had the power "to enquire, hear, and determine" all that pertained to weights and measures law and to punish anyone found guilty of breaking that law. More than two men could be assigned if the shire's size or

---

26. See *Calendar of the Patent Rolls: Henry III: 1266-1272* (London, 1913), pp. 630, 676, 685.

27. Typical of the charters bestowing weights and measures administrative authority on certain citizens in the early fourteenth century was the following: "Commission to Henry de Campo Arnulphi and Reynold de Botereux, reciting that in *Magna Carta* it is contained that there be one measure . . . and that on 20 February, 10 Edward II, on the frequent complaint of the magnates and chiefs at divers Parliaments and of others of the realm that certain merchants and others used divers measures, to wit, larger with which they bought, and smaller with which they sold, to the great deception and manifest loss of the people and contrary to the said charter, the king caused proclamation to be made through every county in the realm . . . prohibiting merchants . . . on pain of forfeiture from buying or selling by other measures than those approved by the standard of London, or from otherwise using them, and afterwards the king caused to be made certain measures for measuring corn, wine and ale, by the said standard, and to be sent to the principal town of each county . . . that certain measures in those counties might be made and proved by the measures so approved. . . . Now the king has understood that certain merchants . . . not fearing the said proclamation and prohibition, use measures disagreeing with the said standards . . . to the grave deception of the people. . . . [The King, therefore, has appointed] the said commissioners to survey all measures whereby wine, ale and corn are sold and bought in the county of Cornwall, and to cause the same to be approved by the said standards and to burn all false measures found by them, and to make inquisition in that county as often as need be touching the names of those who have used other measures than those approved . . . and to punish all such by ransoms and amercements according to their guilt" (*Calendar of the Fine Rolls: Edward II: 1319-1327*, pp. 314-15). The names of all commissioners appointed, contemporaneously, to other English shires may be found in the same source, pp. 315-16. See also *Munimenta gildhallæ Londoniensis*, p. 383.

population density warranted it. In a second clause Parliament authorized the commissioners to retain a fourth part of the fines to cover their expenses and to "answer to the King for the other three parts." The enactment ended by insisting that the new appointees were not to usurp any of the metrological duties of local clerks of the market.

Besides the ambiguity in the first clause of the word "survey" and of the phrase "good and sufficient persons," the second clause was potentially troublesome and it proved within a relatively short time to be the undoing of the entire statute. To begin with, Parliament overlapped the territorial and administrative jurisdictions of weights and measures officials, and would do so repeatedly throughout the medieval and early modern periods. Edward's statute does not say that the commissioners were to assist local clerks of the market or were to ally themselves and their programs with those being conducted presumably by the clerks. It never attempts to divide certain functions or geographical areas among them. If both groups had the same duties and the same regions to patrol, then Parliament was duplicating the work performed by its available manpower and inviting intrashire rivalries among the very people responsible for carrying out its standardization directives. One can see readily how the clerks—an older and more established officer corps—would view the newly arrived commissioners: the latter were interlopers or, what may have been worse, "agents" sent out by London to check up on their work. In either case, resentment quickly turned to hostility and neither group was entirely to blame for the subsequent collapse in weights and measures administration. A way should have been found to prevent the encroachment of one group on another, which could have been accomplished by spelling out in detail their various individual and interrelated functions or by simply eliminating the clerks altogether if they were judged incompetent by the central government.

Furthermore, the commissioners should have been reimbursed according to some mutually agreed upon annual stipend and not according to a pro rata proportion of all fines collected. Parliament's reasoning was that the latter method of payment would encourage officials to be not only diligent but severe in the execution of their commission since their own profit, as well as the Crown's, was at stake. As an ancillary benefit, the snail's pace of past standardization programs would be speeded up considerably. Unfortunately, such an inducement worked not to the interests of the Crown or to metrological uniformity but principally to fattening the pocketbooks of the commissioners. Forced exactions, failure to deliver to the Exchequer the sums collected each year on the Morrow of St. Michael's Feast, and other extralegal and illegal activities led to Parliament's repeal of the commissions in 1344 and to the declaration "that from henceforth no such commission shall go out." London ordered the men appointed over the last four years to appear before the treasurer and the barons of the Exchequer to render an account of what they collected. No one could claim immunity from prosecution. Anyone accusing the commissioners of malfeasance of office was entitled to a hearing and if the evidence supplied was deemed sufficient, then writs were to be issued to local sheriffs enabling the injured parties to seek a redress of grievances in London.

Parliament's decisive action in 1344 in response to the many appeals from merchants and other concerned urban residents was both necessary and laudable. Local patience was rapidly waning and the legislators prevented what was surely to become open rebellion. But why was the situation allowed to grow to such alarming proportions? Even before passage of the 1340 enactment there were frequent appeals to London concerning the misdeeds of commissioners. In 1326, for example, certain citizens of Norfolk complained of being punished summarily by Robert Baynard and Simon de Hedersete—two commissioners recently named to this area—even though, previously, the infractions cited by the commissioners had been brought to the attention of the city's bailiffs and proper fines and amercements had been levied on all guilty parties. Townsmen deplored the fact that they were being punished twice for each and every weights and measures violation. After they certified that their weights and measures now conformed with those recently received from the Crown, the commissioners were instructed "not to intermeddle further with the measures, and not to molest or aggrieve the said [citizens]." If bailiffs had been negligent in punishing any excesses, then Robert and Simon were to report such infractions to the king and he would act in a just and reasonable way to remedy the situation.[28] Similar complaints were commonplace.

Parliament was cognizant of these problems but none of the appointments recorded in the Patent Rolls between 1340 and 1344 warned commissioners in advance to refrain from such activities. In the appointment of Richard Spynes, Edmund de Lacy, and Nicholas Halden to Yorkshire in 1341, the charter merely went one step beyond the statute by insisting that as surveyors of weights and measures they were to perform their duties "without prejudice to the clerk of the market in his office or lords of liberties within the county."[29] One year later a commission to Walter de Henle, the king's sergeant-at-arms, and Nicholas de Banbury for the county of Salop did not even duplicate the statutory provision for

28. *Calendar of the Close Rolls: Edward II: 1323-1327* (London, 1898), pp. 532-33. See also *Select Cases concerning the Law Merchant: A.D. 1239-1633*, pp. xlvii-xlviii.
29. *Calendar of the Patent Rolls: Edward III: 1340-1343* (London, 1900), p. 318.

it made no mention of any possible source of administrative conflict.[30] The same is true in the contracts for William de Chiltenham, Robert Dapetot, William de Colford, and John de Bekyngton for Gloucestershire and for John de Whetlay, Nicholas de Appelby, and Richard de Esshewra for Yorkshire, the first document dating from 1343, the latter from 1344.[31] Appointments of commissioners to 16 other shires are basically identical in wording to these.[32]

The law of 1344 that repealed the commissions lasted seven years. By a statute of 1351 the government lifted its ban and ordered newly appointed commissioners to begin their duties at once. But the role that such men would play thereafter in the urban management of weights and measures was limited principally to inspection and verification procedures, while enforcement duties generally fell to the king's justices, to one or more of the circuit courts, or to elected city officials, notably bailiffs and mayors. Moreover, although one can find evidence of the existence of commissioners in various records after 1351, most of their remaining functions were gradually taken over by the officials mentioned above during the late fourteenth and early fifteenth centuries. As their authority ebbed, commissioners became assistants to mayors, bailiffs, and justices even though there is evidence showing two specific commissions with more than advisory jurisdictions.[33] By the sixteenth century such groups—staffed almost exclusively by scientists and celebrated dignitaries of the central government—concentrated on creating new concepts for metrological legislation and on constructing more precise physical standards.

30. Ibid., p. 388.
31. *Calendar of the Patent Rolls: Edward III: 1343-1345* (London, 1902), pp. 72, 282.
32. See ibid., pp. 282 and 283, for the names of the commissioners and their assigned shires. See also the *York Memorandum Book, Part II,* p. 260; *Rotuli parliamentorum,* 2:141; and *Select Cases concerning the Law Merchant: A.D. 1251-1779,* ed. Hubert Hall, Selden Society Publication, 49 (London, 1932): 161, for additional names and related information.
33. The two commissions were these, dating from 1441 and 1461 respectively: "Each year at the wonted time of the election of the bailiffs, the burgesses may choose four of the better men of the town [of Colchester] to be with the bailiffs . . . for a year from their election justices of the king's peace within the said town and liberty. . . . These four men and the bailiffs or any two of them, shall be justices of the peace therein and have as full powers and authority as other justices of the peace and justices assigned to hear and determine felonies, trespasses and other misdeeds, and justices of weights and measures, huntsmen . . . servants, labourers and craftsmen have in any county or place within the realm, to the exclusion of the jurisdiction of any other justice of the peace" (*Calendar of the Charter Rolls: 5 Henry VI-8 Henry VIII: 1427-1516* [London, 1927], p. 84); "Commission to Henry Brook to inquire touching the use of false weights and measures throughout the realm and to arrest and imprison any offenders, and to certify the treasurer and barons of the Exchequer . . . of the names of the persons found guilty and of the fines imposed upon them" (*Calendar of the Patent Rolls: Edward IV: 1461-1467* [London, 1897], p. 134).

They were now quartered permanently in London and supervised directly by Parliament. They no longer had any functions to perform in urban areas.

Urban Officials

Long before the demise of the commissioners' functions, mayors, bailiffs, and aldermen had become vitally important in the administration of weights and measures law throughout most urban centers in the British Isles. Even though the total number of people living in metropolitan areas was appreciably smaller than the number in the agricultural regions, many more legislative enactments and royal decrees were directed specifically to cities and towns than to villages and hamlets. Part of the explanation for this unusual situation was that since cities had emerged as the principal depots for an expanding commercial and industrial complex, rural areas had become increasingly dependent on them, and rural residents were forced to travel to them more and more to conduct their business transactions. Correspondingly, as the financial base in urban areas improved and enlarged, the central government recognized that here was its single most important source of taxation revenue. The weights and measures of such cities had to be made to conform to Crown standards so that London would not be cheated, either in money or in products, in business dealings with other cities. Significant, too, was the fact that urban areas had developed much later and were not confined so rigidly by custom and tradition. It was virtually impossible for London to alter appreciably the metrological practices of manorial and nonmanorial rural settlements largely because time had fixed certain units and physical standards, and in a domain where custom had long since become law, governmental directives aimed at modification or change, regardless of the moderation implicit in any standardization program, were viewed generally with suspicion, sometimes with hostility. Urban areas, on the contrary, had a multifaceted economy, one where profit was the end-all of the labor force. Where profit rules in lieu of tradition and centuries-old norms, change is viewed more as a required condition of life than as a means of destruction. Furthermore, urban residents were more cosmopolitan and heterogeneous than their rural neighbors. Instead of resisting change, they accepted it and frequently welcomed it. But even they were hostile to standardization programs if their margin of profit—obtained either legally or illegally—was interfered with in any way. Yet they were reacting out of self-interest and self-aggrandizement. They were not preconditioned to oppose the central government out of habit. London found it far easier to achieve metrological uniformity in cities and towns than in their rustic environs. Crown and Parliament concentrated on urban centers, hoping that the success obtained would filter through to the world beyond their walls. As it happened, time, coupled with the commercial, industrial, and agricultural revolutions, took care of the countryside.

There is no government document or public record that provides a clue as to whether city officials had weights and measures administrative responsibilities before the eighteenth year of Edward II's reign. The only mention of them in regard to any facet of weights and measures prior to the fourteenth century is in an entry in the Patent Rolls, dated 1276, instructing the bailiffs and "good men" of Bristol to buy and sell grain solely by the London quarter.[34] This is basically a mandate prohibiting the heaped quarter and nothing in the entry relates to inspection, verification, and enforcement. Presumably the city fathers had not yet acquired such privileges and would not be entitled to them legally until the time of Edward II. It is imperative to note, however, that when they did receive certain administrative functions in the first quarter of the fourteenth century, they got them precisely at the time that local complaints to Parliament concerning the abuses of commissioners began to intensify. Whether there is any causal relationship between the two events must remain conjecture since no source ever credits the rise of the one with the fall of the other. But it is perhaps more than coincidental that they occurred simultaneously.

The acquisition of administrative functions by urban officials was not gained through only one parliamentary enactment or even during the lifetime of one monarch. Edward II's ordinance of 1324, being, perhaps, similar to most other ordinances in that it was an emergency measure made operative before a more complete, statutory rendering could be drafted, merely provided mayors and bailiffs with the right to examine weights and measures twice a year and to verify them through the use of proper seals. They also became custodians of their cities' standard bushels and ells. But in a second clause each city's standards were placed under the supervision of "six lawful persons . . . before whom all measures shall be sealed." As of 1324, thus, mayors and bailiffs could determine only whether weights and measures in their respective areas conformed to statutory specifications. Their verification and custodial duties were shared with the commissioners. The enactment represents, fundamentally, a transition from an earlier period dominated by commissioners to a later one that will see town officials performing all of these administrative duties unilaterally.

Further growth in the jurisdictions of mayors and bailiffs began soon

---

34. *Calendar of the Patent Rolls: Edward I: 1272-1281* (London, 1901), p. 172.

after the death of Edward II. In 1328 Parliament enacted that all cloths imported into the realm had to be measured before distribution or sale by the king's alnagers. In towns where such examinations took place, the alnagers had to have their results verified in the presence of mayors and bailiffs or before bailiffs alone if there were no mayors. These officials verified that the cloths' dimensions conformed to all applicable assizes by placing a special mark on each acceptable fabric. Defective cloths were marked accordingly and forfeited to the king. Each year on the Morrow of St. Michael's Feast, town officials delivered directly to the Exchequer all indentures made on defective cloths. The statute makes it clear that no fees could be collected by mayors and bailiffs for any services rendered.

State and local records of the fourteenth century that postdate the 1328 statute show a substantial increase in the administrative powers of urban officials. In a collection of London memorabilia, for instance, an entry for 1347 finds Geoffrey de Wychingham, the mayor, ordering his sergeant-of-the-chamber to summon before him and his aldermen all of London's turners to answer complaints charging them with "manifold falsities and deceits" in the manufacture of wooden wine and ale measures.[35]

> Upon which Friday came the makers of the said measures. . . . Injunctions were given to them by the said Mayor and Aldermen, in future not to make any such kind of measures of any other wood than dried; and that the measures, when so made by them, must agree with the standard of the Aldermen in whose Ward such measures shall be used, and by the same standard be examined. And that each of these makers should have a mark of his own, and should place such mark upon his measures, on the bottom thereof . . . when by the standard they should have been examined. . . . When any measure made by one of the makers aforesaid shall in any tavern or brewhouse be found to be false or defective, then as well the person by whom such measure shall have been made, as he who shall have sold by such measure, shall incur the punishment at the discretion of the Mayor and Aldermen.[36]

Besides Wychingham's ability to call the entire membership of a guild before him when he deemed it expedient, this is the earliest indication of any mayor or aldermen exercising the powers of enforcement. Three years later a successor to Wychingham, Walter Turk, punished one John de Hiltone, a member of the London Pewterers' Guild, for making pewter

---

35. In the Middle Ages a turner (spelled "turnour" in the cited document) was a craftsman who fashioned, principally from wood, all types of drinking vessels, casks, various tools, and certain other articles, by turning them on a lathe. Here it refers specifically to makers of liquid-capacity measures.

36. *Memorials of London*, p. 235.

measures with a much greater proportion of lead to tin than was allowed by law, "in deceit of the people, and to the disgrace of the whole trade."[37] After a jury of the wardens of this craft substantiated the charges levied against Hiltone, Turk and his aldermen ruled that the measures in question were fraudulent and they were forfeited "to the use of the commonalty."[38]

In yet another and far more significant case, a mandate was sent from Parliament in 1351 to the mayors, bailiffs, and other public officials in Ireland ordering them to insure "that the same assize of measures and weights as the king uses in England be observed in all [Irish] cities, boroughs, market towns and other places . . . and [to make] public proclamation in his name that none in buying or selling use other measures and weights . . . under pain of heavy forfeiture."[39] All Irish officials were further instructed to attend, counsel, and help Robert de Chaundos, the king's yeoman, "to whom the king had committed the office of measures in Ireland."[40] Twenty years later the mayor and bailiffs of Waterford had procured so many of the functions exercised normally by the local clerk of the market that Parliament prohibited any clerk from entering the city "to do anything related to his office . . . [save] only to supervise, approve, and examine the standard of the city and to correct defects or excesses of the same."[41] All profits accruing from the administration of weights and measures, except for those associated with the unspecified standard, belonged to the mayor and bailiffs.

By 1371 the mayor and chamberlains of York had acquired the same privileges accorded the Irish urban officials even though no reference was made in any of the documents to York's clerk of the market.[42] The identical situation occurred at Rochester, in Kent, in 1376. In this town it was the constables and bailiffs who held administrative duties since there was no mayor.[43] Other town officials would be awarded similar privileges during the remainder of this century and during the next.[44] In the Tudor era they would reach the zenith of their metrological authority.

37. The commonly accepted ratio was 16 pounds of lead to every hundredweight (112 pounds) of tin.
38. *Memorials of London,* pp. 259-60.
39. *Calendar of the Patent Rolls: Edward III: 1350-1354* (London, 1907), p. 123.
40. Ibid.
41. *Calendar of the Charter Rolls: 15 Edward III-5 Henry V: 1341-1417* (London, 1916), p. 219.
42. *York Memorandum Book, Part I,* p. 15.
43. *Rotuli parliamentorum,* 2:349.
44. See especially the detailed accounts in *Calendar of the Charter Rolls: 15 Edward III-5 Henry V,* pp. 297-98; *Monumenta Juridica: The Black Book of the Admiralty,* ed. Sir Travers Twiss (London, 1873), pp. 176-77; and *Rotuli parliamentorum,* 3:272, 323.

The Rural Aristocracy

Throughout the Middle Ages the aristocracy exercised a virtual monopoly over weights and measures in rural areas. The great magnates almost always received metrological duties when they acquired estates from their king. They, in turn, frequently bestowed similar functions on their vassals when they subinfeudated portions of their estates. Manorial courts were concerned with investigating presentments dealing with the use of unfair measures by the serfs and with levying an assortment of fines for transgressions of manorial standards. Sometimes bailiffs directed all proceedings; sometimes they merely issued precepts to stewards; sometimes stewards controlled all of the major functions; sometimes all of these manorial officials worked with or for justices of the peace. But the power remained in the hands of the aristocracy in many places until as late as the nineteenth century.

When one considers that the vast majority of people in the British Isles lived under one form or another of manorialism and feudalism throughout most of the Middle Ages, and that even after the demise of those institutions many people remained tied in some way to a rural aristocracy, it is surprising that records describing the weights and measures functions of lords are not more abundant. For example, there are hundreds of entries in the Close Rolls referring either to the acquisition by lords of control over weights and measures or else reaffirming the lords' traditional jurisdiction over such matters, but in only one case are these duties ever described. In an entry dated 1331, Edward III granted to Ebulo Lesfrange, Alesia, his wife, and his heirs

> custody and ward of the castle of Lincoln with the bailey . . . royalties, liberties and free customs and all things pertaining to [them] . . . as fully as Henry de Lacy, late earl of Lincoln held them. . . . Also . . . the assay of measures, to wit of bushels, gallons, pottels, quarts, . . . ells, and other measures of corn, wine, and ale, and the assay of weights, to wit of pounds, stones, and other weights of all things sold by weight, and forfeitures of measures and weights . . . whenever they were found to be false, with the amercements and other punishments for the same.[45]

Ebulo could inspect, verify, and enforce. But even in this document one is supplied only with data concerning *what* he could do; nothing therein relates to the methods he was expected to use. This situation is indicative, perhaps, of the position of weights and measures administration in the feudal-manorial world. Custom was supposed to dictate "how" something was to be carried out, while grantors decided "what" would be given or performed.

---

45. *Calendar of the Close Rolls: Edward III: 1330-1333* (London, 1898), p. 255.

A similar case in point occurred in 1351 when the residents of Chester complained that millers kept "divers measures which are not fairly or reasonably made." The Black Prince gave orders to his vassals that the measures must be examined and amended. After repairs they were to be "marked with the mark of the prince's tillerie," which the chamberlain kept in his possession. In the autumn of that year the justice and chamberlain recommended to the prince that all kinds of weights and measures used in markets should be standardized throughout Cheshire. The prince agreed and ordered that the standards be kept at Chester Castle.[46] He personally decided in each instance what would be done, but nothing is said as to how it was done.

Such information may have been left out of these accounts for one particular reason. If the actual procedures were decided ordinarily by the vassal, there would be little purpose in cataloging them in any charter. A grantor would have expected a recipient to handle the administration directly or to delegate some or all of the functions to his manorial court. This the latter often did, and since recent scholarship has proved that no two manors ever operated in exactly the same way, one should not expect to find any standardized procedures in matters relating to weights and measures administration. As long as the law was enforced and malfactors were properly punished, the grantor was satisfied. He never expected—or, more correctly, could not expect—more than this.

Church, University, and Royal Dignitaries

No commissioner, mayor, bailiff, alderman, or manorial lord was a permanent, full-time weights and measures administrator. Nor was any church, university, or royal official. None could be and still perform his business or professional duties competently. Monarchical and parliamentary directives notwithstanding, personal matters always outweighed other considerations. The metrological functions of every one of these people was of secondary or tertiary importance to their political, mercantile, industrial, military, or religious interests. Under such circumstances it is understandable that kings and national legislatures labored long and repetitiously for reform. Statutory wording was rarely precise and physical standards certainly were not always reliable, but the results obtained would have been more substantive if administrative duties had not been delegated so liberally to so many part-time employees.

Like urban and rural residents, many regular and secular clergymen, certain university chancellors, and the chamberlains and Exchequer barons of the central government were given inspection, verification, and

46. H. J. Hewitt, *Medieval Cheshire: An Economic and Social History of Cheshire in the Reigns of the Three Edwards,* Chetham Society Publication, 88 (Manchester, 1929): 187.

enforcement duties at various times during the Middle Ages. There were numerous reasons for these grants but four seem to have been of cardinal importance. First, metrological duties were bestowed traditionally on citizens in recognition of services rendered to the king or Parliament. In this case the power to administer the laws was presented in lieu of or in addition to some other honor or reward. Second, there was occasionally a national or a local emergency at hand and the capital needed more than the normal number of public servants to bring the situation to a peaceful and abrupt conclusion. When the issue was finally resolved it was usually impossible to get any of the benefactors to relinquish their newly acquired rights. Third, when the Crown or Parliament issued privileges on a piecemeal basis to a select few, an avalanche of similar requests from other interested parties invariably followed and the government had to appease them. Besides the many public-spirited sentiments, grantees were motivated by sundry political and economic considerations, by jealousy, or by an urgent desire to maintain themselves on an equal footing with their peers. Finally, metrological duties were often included in charters either as ancillary benefits or as part of a long list of regalian rights presented by some monarch to local dignitaries whose support he needed. Whatever the reason, these practices mushroomed, and London, Edinburgh, and Dublin soon had an oversupply of untrained weights and measures officials who operated generally outside the ordinary channels of governmental supervision.

The Charter, Close, and Justiciary Rolls are replete with references to acquisition by monasteries of control over weights and measures functions. In one instance the abbot and monks of Evesham were given in 1236 the assize of weights and measures and were asked to cooperate with the city administration to end certain abuses committed by citizens in the buying and selling of merchandise.[47] Whenever a document specifically mentioned "assize," all three major functions were inferred: the religious community could examine, verify all legal weights and measures, and punish those perpetrating frauds. Most of the grants in the Rolls are worded in this fashion. In another entry dated 1348, an Irish abbot and his monks are provided with standards "and all other things belonging to the office of the market" so that "in the absence of the king or his chief justice they shall have . . . custody and essay of weights and measures in their lands and fees with the correction of defects . . . and shall carry on the same by their ministers, as lords of liberties in the said land are wont to do . . . provided always that the exercise hereof be had reasonably and justly for the common profit of the king's people."[48] An English abbot and

---

47. *Close Rolls of the Reign of Henry III: 1234-1237* (London, 1908), p. 289.
48. *Calendar of the Charter Rolls: 15 Edward III-5 Henry V*, pp. 98-99.

convent in 1446 received the assize of weights and measures and "all else belonging to the office of the clerk of the market of the king's household, with power of punishment as fully as the said clerk would have had before this grant; and they shall have the amercements, fines, and other profits thence arising to the exclusion of the jurisdiction of the said clerk."[49] Two years later an abbess and the nuns of her convent were given the same privileges as well as "the assay, amends and assize of bread, wine and ale and all other victuals."[50] The remainder of the entries in the Rolls are similar in form and wording to these.

Similar grants often went to bishops and archbishops. The following three citations, dated 1300, 1394, and 1463 respectively, are typical of the vast majority found throughout the Rolls. Note that the first is Irish in origin.

> The Archbishop comes and says that he does not claim to hold all pleas of the Crown, but only the following:—the shedding of blood of Englishmen, *Vetitum namium,* assize of bread and ale, measures and weights, hue and cry unjustly raised, waif and stray of animals unclaimed within the year, and purpresture made on him or others in his fee and this from ancient time.[51]

> Grant of special grace and out of devotion to St. Thomas of Hereford and affection for John, bishop of Hereford . . . to the bishop . . . of the assize of bread, wine and ale and all other victuals and the assay and sealing . . . of all measures and weights within . . . the city and suburbs of Hereford, and the cognisance, punishment and amends of bakers, brewers, taverners, victuallers, regrators, forestallers and other such transgressors within their fee and liberty . . . so that the clerk of the market of the king's household shall not enter the said . . . [city] to do any of the foregoing or aught else touching his office nor intermeddle therein in any way.[52]

> The archbishop shall have . . . assize of bread, wine and ale and other victuals and of measures and weights and other matters pertaining to the office of the clerk of the market . . . with the punishment of the same and all else thereto belonging . . . and the amercement, fines and profits thence arising . . . to the exclusion of the clerk.[53]

In 1327 the chancellor of the University of Oxford became the first member of his profession to be named a weights and measures administrator. As recorded in the Patent Rolls, this most esteemed educator was responsible for enforcing the assize of weights and measures throughout the town and its suburbs. He, personally, had to supervise inspections,

---

49. *Calendar of the Charter Rolls: 5 Henry VI-8 Henry VIII,* p. 58.
50. Ibid., p. 92.
51. *Calendar of the Justiciary Rolls or Proceedings in the Court of the Justiciar of Ireland,* ed. James Mills (London, 1914), 1:316.
52. *Calendar of the Charter Rolls: 15 Edward III-5 Henry V,* p. 349.
53. *Calendar of the Charter Rolls: 5 Henry VI-8 Henry VIII,* pp. 193-94.

order the destruction of any fraudulent weights and measures, supply lawful ones in their place, and inflict punishment on those who broke the law.[54]

That these duties were executed precisely and thoroughly seems to be evident from later entries in the Rolls, in particular from one in 1346, which describes how certain urban officials resented the chancellor's new position and his stringent interpretation of the law. A serious debate ensued which threatened to blow up into a major confrontation. At the center of the controversy was the same problem that created friction between commissioners and mayors: a duplication of metrological functions. In this case, the city fathers had received much earlier the right to inspect, verify, and enforce Oxford's weights and measures twice every year. When the chancellor sought the cooperation of the mayor in performing his new tasks, the mayor refused, largely due to the intransigence of his aldermen. The Crown ordered the mayor to stop vacillating and to "act with the chancellor in keeping the . . . assize." The aldermen and burgesses were reprimanded and enjoined to assist them. If anyone disobeyed this warning, the king threatened "to take into his hands their liberties and inflict such further punishment on them as shall serve as a deterrent to others."[55] Nothing more was ever reported concerning the matter.

What Edward III did for Oxford, Richard II would accomplish for Cambridge. The latter case, however, was different in two respects. First, the grant given Cambridge had a time limitation. At a parliament held at Gloucester in 1378, the chancellor received the assize of weights and measures but only for a five-year period. Second, it was intended to punish the mayor and bailiffs, whom citizens had accused of gross negligence in the enforcement of weights and measures law. The chancellor, assisted by the scholars of the university, was empowered to punish the city fathers if further complaints were made.[56] By 1381 Crown and Parliament believed that the mayor and bailiffs had been chastised sufficiently, and they were permitted to resume their metrological functions. At the same time the chancellor was awarded the same grant given his Oxford counterpart in 1327. He and his scholars were to cooperate with urban officials in enforcing the law.[57]

54. *Calendar of the Patent Rolls: Edward III: 1327-1330* (London, 1891), p. 21.
55. *Calendar of the Patent Rolls: Edward III: 1345-1348* (London, 1903), pp. 102-3.
56. *An Exact Abridgement of the Records in the Tower of London*, ed. Sir Robert Cotton (London, 1657), p. 172; and *Rotuli parliamentorum*, 3:68.
57. The grant reads: "Le Roi . . . fist doner & comittre as Chanceller & Escolers de la dite Universitee la garde de l'Assise de payn, vin & cervoise, & la conissance & punissementz d'icelles; & auxint la garde de l'Assise & de l'Assaie & la Surveue des mesures & poys en la dite Ville, & les Suburbes d'icelles" ["The king . . . has given and committed to the

The only other instance in which an educational institution received such privileges occurred in 1587, when the Scottish legislature granted to the chancellor of the College of Glasgow "*the* custumes of *the* tron wechtis and of all merchandice and geir vsit to be weyit at *the* said tron w*ith* all mesouris grite and small vsit within *the* said citie of Glasgow."[58]

In an act of James III made in Edinburgh in 1467, the chamberlain was entrusted with the overall supervision of Scotland's weights and measures. He, together with the sheriffs, was responsible mainly for enforcing statute law. Before the passage of this enactment, however, the chamberlain's court already had jurisdiction over the inspection and regulation of many matters connected with commerce. They frequently made trips throughout Scotland, carrying with them copies of the standards in order to conduct inspections and verifications of the weights and measures in the burghs. They also appointed inspectors who examined and certified the quality and quantity of cloth, bread, and casks for alcoholic beverages.

The English chamberlain never acquired so wide a range of duties but rather assisted the monarch in disseminating weights and measures standards throughout England. He served also as a special agent whose responsibilities varied according to the emergency at hand. He rarely inspected or verified weights and measures and there is no indication that he ever enforced directly any aspect of the law.

Much the same can be said for the barons of the Exchequer, who were important primarily as guardians of the Exchequer standards. Occasionally they supervised inspections of those weights and measures brought before their court in London for comparison with the Crown standards. The following is a typical example.

> On May 12 . . . comes before the Barons a certain J. Wyldgrys of Coventry in his own person and showed the Court here two iron rods, one a yard long, another an ell long, used for measuring saleable cloth and other measurable things to be sold in the city, begging through the favour of the court in the city's name that these rods be placed by the side of the standard of the Exchequer, to prove if the same yard and ell were legal measures and corresponded to the Exchequer standard or no. Whereupon on the same day the said yard and ell having been compared with the Exchequer yard, it appears to the barons that the said yard brought by J. Wyldegrys is of standard length and that the ell is equal to the length of a yard and a quarter

---

chancellor and scholars of the said university the custody of the assize of bread, wine and ale, and the cognizance and punishment thereof; and also the custody of the assize and the assay and the survey of weights and measures in the said city and its suburbs"] (*Rotuli parliamentorum*, 3:109).

58. *The Acts of the Parliaments of Scotland: A.D. MDLXVII-MDXCII*, Great Britain Record Commission Publications (London, 1814), p. 487.

according to the standard yard of the Exchequer. The same yard and ell thereafter, that is to say on May 20, 1474, are delivered sealed with the Exchequer seal to J. Wyldegrys . . . to be used as lawful measures.[59]

Coroners, Sheriffs, and Justices

Unlike the officials described above, coroners, sheriffs, and justices were permanent weights and measures administrators. That is, the inspection and verification of weights and measures, together with the enforcement of statutory law, were performed regularly by them as part of their professional responsibilities. Such functions came with their jobs. They were not reserved for only a select few within the group. They were not one aspect of a hastily conceived plan to bring some local emergency to a rapid termination. They were not special favors or bonuses awarded for previous meritorious service. They were not rigidly confined to the operations of certain trades or businesses. They were not sought after for political, social, or economic status. They were ordinary, not extraordinary, duties, representing a continuous and unending service owed to king and Parliament. They began when coroners, sheriffs, and justices accepted their appointments and ended when they retired, were dismissed, or died. They were not the only, or even the most important, duties exercised by these officials but they were the cornerstone of weights and measures administration for nearly half a millennium.

Elected by county courts, coroners were responsible chiefly for keeping records of all crimes involving Crown rights that occurred in their respective shires. These country gentlemen, generally holding a political and social ranking slightly below that of the sheriffs, were professionally independent of the latter. Serving as a check upon the powers of sheriffs, they were required to be present at the execution of thieves in private courts, at outlawries and at appeals of crimes in the county courts, and at presentments of crimes whether in towns or county courts. In emergencies, such as at the news of sudden deaths, they had to proceed to the scene of such events and hold an inquest consisting of men chosen from the four nearest townships. All findings were recorded immediately in their rolls. The coroners' lands and heirs could be seized if such information was not available at sessions of the general eyre. From 1194, when the office was created by the government of Richard I, coroners quickly assumed weights and measures functions and certainly by the middle of the thirteenth century such duties had become mandatory.

It is unfortunate that only one source offers any complete description of the exact methods employed by these officials in executing their

---

59. *Coventry Leet Book*, pp. 394–95.

administrative functions. But that source is considered an accurate and a thorough one. According to Fleta, writing sometime in the 1270s, a coroner went periodically to all townships and markets within his territory and assembled local bailiffs or other chief officers at a meeting. He instructed them to send him six law-abiding citizens from the township. After swearing this panel in, the coroner instructed them to make an honest collection of weights and measures. In the meantime he appointed a jury of twelve local residents who had to swear that they would answer all questions faithfully. When the weights and measures were collected together, the coroner had to determine whether the lord of the manor at whose court site the inquest usually took place had sufficient instruments of punishment. The minimum requirements were met if the lord had gallows, the pillory, or the tumbrel.

The coroner then asked his twelve handpicked jurors how often and in what cases their lord had formerly taken fines and amercements for the breaking of any assize. They were also to tell him what was paid and who the transgressors were. After determining whether any further infractions had occurred since the last inquest, and after examining the weights and measures collected, the coroner assessed the penalties. Guilty parties who had no previous records, Fleta insists, were treated leniently, but anyone who committed a third offense paid a triple amercement, and those who were sentenced for a fourth time received corporal punishment. Nothing is said concerning penalties for second violations. His final duty was to order all illegal measures destroyed and before he departed he saw to it that the punishments were carried out.[60]

Sheriffs seldom inspected or verified weights and measures. They were, above all else, law enforcement officers. Assisted by under-sheriffs, itinerant sergeants, hundreds' sergeants, beadles, sub-beadles, bailiffs, and clerks, and appointed directly by the Exchequer, they were responsible for conveying and executing all writs, the procedures of which varied according to the orders contained in them. In addition, they frequently levied distresses, collected dues, summoned juries, and attached persons to appear in court.

As early as the reign of Henry II, sheriffs were entrusted with delivering measuring rods and weights to urban magistrates throughout the realm. Yet it was not until the fourteenth century that documentation is sufficient to afford overwhelming proof of their indispensability to the central government. In 1317, for example, the sheriff of Lincoln was ordered to "cause proclamation to be made that no merchant or other

---

60. *Fleta,* ed. H. G. Richardson and G. O. Sayles, Selden Society Publication (London, 1955), pp. 118–22.

shall, under pain of grievous forfeiture, use any other measures to buy or sell with than those approved by the standards of London, causing it to be known that the king will shortly send certain persons throughout the realm to examine measures and to punish delinquents in this respect."[61] All sheriffs received balances and weights for weighing wool in 1354 and they had to proclaim a general examination of all similar apparatus used by citizens in their territories. The same year saw the sheriff of Southampton posting notice that weights and measures must be brought to Winchester Castle for comparison with the standards kept there.[62] Two years later a detailed order to Yorkshire's sheriff read:

> Whereas a statute . . . contains that all measures, to wit the bushel, half-bushel and peck, gallon, pottle and quart, shall agree with the king's standard . . . the quarter to contain 8 bushels by standard and no more . . . each measure of corn shall be stricken and not heaped saving the rents and ferms of lords which shall be measured . . . by the same measure as was customary heretofore, . . . the king has caused certain measures to be made agreeing with the standard, which he sends to the sheriff to remain with him. . . . Order to receive those measures and immediately to cause proclamation to be made in that county, in market towns and other places, that all having such measures shall bring them . . . and have them made in conformity with the standard.[63]

In 1357 the sheriff of Wiltshire was commanded to take the standards recently issued to him by the treasurer and hand them over to three justices who had just been appointed to assay the shire's weights and measures. After the examination was completed, the sheriff had to reclaim the standards and see to it that they were kept in perfect condition until the next assay.[64] Several months later the sheriff of Yorkshire was given the same instructions with one major change: he had to supervise the construction of copies modeled after the standards and then turn the copies over to the justices for use in the assays. The original standards were never to leave his possession.[65]

Many more illustrations could be provided for the fourteenth and later centuries. They all attest to one fact: sheriffs were metrological "lawmen" who insured nationwide compliance with London's mandates. Other officials, such as commissioners, justices, and clerks of the market, almost always performed the technical operations associated with weights and measures administration.

61. *Calendar of the Close Rolls: Edward II: 1313–1318* (London, 1893), p. 455.
62. *Calendar of the Close Rolls: Edward III: 1354–1360*, pp. 101–2.
63. Ibid., pp. 162–63. See also pp. 183 and 226 for similar orders to the sheriffs of Oxfordshire, Berkshire, and Devonshire.
64. Ibid., p. 365.
65. Ibid., p. 376.

During the thirteenth and fourteenth centuries, justices of assize and of oyer and terminer investigated infractions of weights and measures law and punished transgressors. In the earliest reference to such activities, assize justices were sent throughout England in 1256 to hold examinations for the purpose of determining whether local weights and measures conformed to Crown standards, to eradicate sundry illegal weighing and measuring practices, and to punish those found guilty of crimes.[66] In 1341 two justices of oyer and terminer were assigned to each county to supervise the weighing of sacks of wool destined for shipment to London and to insure that the statutory ruling allowing 14 pounds to the stone and 26 stone to each sack was obeyed by sellers and buyers.[67] The justices' principal concern was to prevent London from being cheated either through the short-weighting of such merchandise by merchants or through the acceptance by receivers of kickbacks from merchants for cooperating in illegal transactions. A Gloucester roll of 1343 describes how the king, ever desirous of maintaining uniform standards throughout the realm, had appointed justices "AD INQUIRENDUM ET SUPERVIDENDUM IN COMITATU GLOUCESTRESIRE, TAM INFRA LIBERTATES QUAM EXTRA, QUOD MENSURE ET PONDERA SINT STANDARDIS REGIS CONCORDANCIA" ("to inquire and survey in the county of Gloucestershire, as well within liberties as without, that measures and weights be in accordance with the king's standards").[68] Two entries in the Rolls of Parliament, dated 1351 and 1355, stipulate that justices were to try each infraction separately so that injured parties and the king received the maximum monetary settlements due them.[69] In the later document, Edward III prohibited justices of laborers, their deputies, viscounts, coroners, and their agents from interfering in the examinations, trials, and other proceedings conducted by his oyer and terminer personnel.

Even before the 1355 grant, justices of oyer and terminer had acquired the right to regulate all weights, measures, and balances employed at trade fairs, concentrating chiefly on the elimination of illegal weighing and measuring practices. They were authorized to hold sessions where litigants could plead redress of grievances against merchants and other businessmen. Fines were generally fixed at four times the value of the merchandise sold and prison terms of one year could be awarded as additional and maximum punishments.[70]

66. "Annales de Burton," p. 375. An identical situation is in the *Rotuli parliamentorum ut et petitiones, et placita in parliamento tempore, Edward R. I.,* ed. Rev. John Strachey et al., 1 (London, 1832): 308.
67. *Rotuli parliamentorum,* 2:133.
68. *Select Cases concerning the Law Merchant (1251–1779),* p. 161.
69. *Rotuli parliamentorum,* 2:240, 265.
70. Ibid., pp. 248–49.

The Calendar of the Patent Rolls lists the appointments of these justices between 1354 and 1358.[71] In most instances they acquired a wide range of duties outside the normal weights and measures sector.

> Commission of oyer and terminer to John de la Lee, Richard de Ravenser, Thomas de Ingelby, Peter de Richemond, William de Nessefeld, Richard Poutrel, and William Warener, in the counties of Nottingham, Derby, York, Westmoreland, Cumberland and Northumberland, touching all trespasses in Queen Philippa's forests, woods, parks, chaces and warrens and all other trespasses, conspiracies, oppressions, extortions, confederacies, champerties, embraceries, forestalleries, falsities, damages, grievances and excesses in her liberties and lordships . . . and those who have used measures and weights not in accord with the standard.[72]

Perhaps no public officials other than clerks of the market ever played a more prominent role in medieval weights and measures administration than justices of the peace. Created by Richard I's government, the office was staffed initially by knights who took oaths to preserve the peace and to assist sheriffs in their police work. Under Henry III their authority grew considerably and such crimes as homicide, treason, incendiarism, robbery, burglary, extortion, and the bearing of arms without license were investigated regularly. By the Edwardian period they enforced laws respecting felonies, trespasses, conspiracies, confederacies, riots, profanations, drunkenness, frauds, unlawful meetings, forcible entries, and disturbances of the peace. They provided for watches along seacoasts and highways and were conservators of the rivers. They inquired concerning those suspected of heresies and those accused of participating in illegal games such as tennis, football, coits, and dice. They even assisted in the execution of the law regulating the apparel worn by various classes of society. In each case they had full authority to punish offenders.[73]

Before 1307 justices of the peace (or conservators of the peace, as they were called until the middle of the fourteenth century) were not always residents of the counties assigned them, but after this date such residence was obligatory. By 1308 they could appoint commissions to carry out their directives and they could pursue offenders from shire to shire, fining and

---

71. In particular see *Calendar of the Patent Rolls: Edward III: 1354–1358* (London, 1909), pp. 121, 256, 396, 549-51.

72. *Calendar of the Patent Rolls: Edward III: 1358–1361* (London, 1911), p. 323. A similar case may be found in the *Calendar of the Close Rolls: Edward III: 1354–1360*, p. 615.

73. For other duties performed by the justices, see T. Nourse, *Campania Fœlix: or, A Discourse of the Benefits and Improvements of Husbandy* (London, 1700), p. 255; W. Roger Breed, *The Weights and Measures Act: 1963* (London, 1964), p. 17; and Charles Austin Beard, *The Office of Justice of the Peace in England: In Its Origin and Development* (New York, 1904), pp. 21, 67-70.

punishing them at their own discretion. But at the same time they were forced to make monthly reports to the council at Westminster concerning their proceedings and they had to include the names of all persons arrested.

By the latter half of the fourteenth century, justices of the peace had become a powerful and effective arm of the central government—a permanent police and administrative institution. They accomplished all of their work at three types of court sessions. The first, labeled discretionary, was an assembly of two or more justices called at a certain time and place to discuss new operations referred to their jurisdiction by statutes and royal commissions. The general or quarter session was a second type and was held four times a year on days prescribed in various enactments. The dates changed repeatedly throughout the Middle Ages. During the fifteenth century quarter sessions met in the first week after St. Michael; the Epiphany; the Clause of Easter; and the Translation of St. Thomas, the Martyr. Each session was supposed to last for three days, but this was ordinarily not adhered to. As a rule quarter and discretionary sessions were held for the entire county, although there was considerable variation in practice. Lastly, special and petty sessions could be held in a given locality at the option of any two justices. A special session investigated offenses with the aid of a jury recruited from the hundred. The chief work of the petty session was to examine the misdeeds of artificers, servants, laborers, vagabonds, rogues, beggars, and paupers and to punish petty thieves and other minor lawbreakers.[74]

No state or local source credits justices of the peace with the performance of any metrological functions before 1360. Whether that date is significant for the formal inception of such duties, or whether it is merely the first testament to their existence, is uncertain. Whatever the case, Parliament in an act of 1360 permitted justices to "inquire, hear, and determine" cases involving violations of weights and measures law and to fine and sentence offenders according to the severity of the infractions. In every shire justices had until the following Easter to explain to citizens the current state of the law regulating weights and measures. Similar instructions were directed to town magistrates.

Only two other references can be found for the remainder of the century. In 1379 a recent appointee to the office of justice of the peace in Kent—one Thomas de Garwenton—was removed from his job "for particular causes" relating to his enforcement of the statutes of laborers

---

74. Actual cases and the names of some of the justices may be found in Beard, pp. 70 ff., and in Richard Carew, *Carew's Survey of Cornwall; to Which Are Added, Notes Illustrative of Its History and Antiquities,* ed. Thomas Tonkin and Francis Lord de Dustanville (London, 1811), pp. 221 ff.

and of weights and measures. He was ordered "to meddle no further therein."[75] Ten years later justices of the peace throughout England were given authority to inquire and hold pleas concerning all aspects of weights and measures law and to punish offenders on behalf of both injured parties and the king.[76] This increased appreciably their already extensive powers. Thomas' alleged misdeeds must have been the exception rather than the rule.

Parliamentary legislation dealing with justices of the peace was unusually prolific during the fifteenth century. A statute of 1413 repeated the provisions found in the 1389 document. In 1417 justices were instructed to inquire into falsification of weights and measures and to imprison offenders without mainprise until they were sentenced or acquitted. If found guilty, defendants were sent to prison until they paid the fines or ransoms levied by the justices. The latter also had authority to hold hearings on weights and measures as often as they thought necessary. In 1423 they were given special duties of "inquiring, hearing, and determining" cases that involved tuns, pipes, tertians, and hogsheads of wine; herring and eel barrels; kilderkins, tertians, and firkins; and salmon butts. Six years later justices and urban officials were empowered to look into alleged violations committed by merchants when weighing by the balance. An act of 1433 ordered justices once again to enforce the statutes and ordinances dealing with weights and measures. They were to try anyone who sold merchandise with illegal measures and to punish guilty parties. The provisions of the acts of 1413 and 1429 were proclaimed in London and all other cities, boroughs, and towns of England. In 1453 Parliament, as a remedy for various abuses committed by guildsmen, compelled all fraternities and incorporated companies to place their letters patent and charters on record before the justices of the peace or chief governors of the towns, and forbade masters, wardens, and others to enforce questionable ordinances until they had been discussed and approved by the justices. The justices and chief governors were further authorized to repeal or revoke any ordinance that they considered unlawful or unreasonable. A statute of 1490 confirmed the powers vested in these justices to investigate infractions committed by merchants and producers in the buying and selling of merchandise. The last act of this century was passed in 1494. After confirming the 1490 enactment, it stated that once justices had determined certain weights and measures to be fraudulent, they were to fine and punish transgressors at their discretion. They could, by examination and inquiry, hear and determine the defaults of all town officers and of buyers and sellers.

75. *Calendar of the Close Rolls: Richard II: 1377-1381* (London, 1914), p. 326.
76. *Rotuli parliamentorum,* 3:272.

All of these duties and responsibilities given justices of the peace continued in effect during the course of the sixteenth, seventeenth, and eighteenth centuries. Thereafter, their metrological authority waned as weights and measures control passed to government inspectors.

The Specialists

Throughout the history of medieval weights and measures administration, only alnagers, port measurers and weighers, and clerks of the market specialized exclusively in performing one or more metrological functions. Arriving rather late on the administrative scene—the first group dates from the thirteenth century and the second and third from the fourteenth—these highly trained, full-time, permanent officials were, normally, never given duties other than those of inspecting, verifying, and enforcing. Receiving their appointments from departments of the central government, they owed their loyalty and service to Crown and Parliament. Local political, social, mercantile, military, religious, and other matters were not their responsibility. They were not answerable to any group, trade, business, or corporate unit for actions or decisions in accord with statute law. Their duties were stated clearly and precisely. It was the exception rather than the rule for the capital to give them emergency powers. Their positions required them to conduct operations on a regularly scheduled basis, either for a limited term or for life. Generally speaking, they were the most proficient corps of metrological administrators prior to the development of the inspectorate in the nineteenth century.

Alnage, or the assize by the king's alnagers of manufactured cloth, dates back at least to the time of Edward I, although no mention of it in a state enactment occurs before the reign of Edward III.[77] Throughout the Middle Ages alnagers performed two principal services: they enforced laws governing cloth measurements and they collected subsidies levied on cloth sales. Their duties extended to examining foreign cloths, marking

77. See *The Letter Books of Joseph Holroyd (cloth-factor) and Sam Hill (clothier); Documents Illustrating the Organisation of the Yorkshire Textile Industry in the Early 18th Century,* ed. Herbert Heaton (Halifax, 1914), p. 13; John Cowell, *The Interpreter: or Booke Containing the Signification of Words* (Cambridge, 1607), sv *aulnegeowr;* Richard Rolt, *A New Dictionary of Trade and Commerce* (London, 1756), sv *alnager;* Lord Chief Justice Hale, "A Treatise, in Three Parts. Pars Prima. De Jure Maris et Brachiorum ejusdem. Pars Secunda. De Portibus Maris. Pars Tertia. Concerning the Customs of Goods Imported and Exported," in *A Collection of Tracts Relative to the Law of England, from Manuscripts, Now First Edited,* ed. Francis Hargrave (Dublin, 1787), 1:239; Lewes Roberts, *The Merchants Map of Commerce* (London, 1677), p. 44; and S. William Beck, *The Draper's Dictionary: A Manual of Textile Fabrics, Their History and Applications* (London, 1882), sv *alnage.* The only source that mentions the performance of alnage duties before the reign of Edward I is the *Rotuli parliamentorum,* 2:230. Here such functions extend back to Henry III's time.

and sealing cloths of assize, and requisitioning defective pieces or bolts on behalf of the king. Appointed and remunerated by the Exchequer, they made certain that buyers were not deceived by sellers. As clerks of the market were appointed to prevent deceits in weights and measures, so alnagers performed a similar service in regard to textiles. By the sixteenth century, however, they were forced to relinquish their searching and measuring duties and to devote all of their time to collecting fees and fines. The cloth industry had grown too large for one set of officials to handle all of these operations. They became part of the machinery of taxation, while newly appointed personnel, called searchers and measurers, acquired their other functions.

Statutory regulation of alnage duties began in 1328. In the presence of mayors and bailiffs, the king's alnager and the latter's deputies were to measure white and colored cloths with cords of 28 and 26 ells respectively. If cloths met these desired lengths and if their widths measured 6½ quarters, they were marked by the cloth officials and bailiffs. Fabrics that did not conform to these specifications were seized and turned over to the king. There is no indication in the document as to what types of verification seals were used.

The 1328 enactment was amended considerably in 1351 when Parliament ruled that all types of cloth sold in England be measured by the alnager and his deputies. Unfortunately, the statute never states what the dimensions of various kinds of cloths should be. It merely says that the alnager must measure cloth by the standard ell. Neither does it reaffirm the continued cooperation among cloth officials, mayors, and bailiffs in the examination and verification procedures. Apparently not all aspects of the earlier act had been carried out diligently since the major emphasis in this act was on what would be done if alnagers and deputies committed frauds in the course of their duties. Only addressing itself to one specific violation, Parliament decided that in cases of default, alnage officials must be tried at the site of their violations by the highest-ranking fair or town officers or by someone else appointed by the king. If they were found guilty, prison terms of one year were meted out. Upon completion of their sentences, such persons were forbidden thereafter to perform any alnage duties. The alnager was responsible for his deputies' actions although the statute does not specify whether this implied personal punishment of the master for the misdeeds of his assistants.

During the next forty years cloth statutes were concerned with establishing standard dimensions for various kinds of fabrics, the most notable being those enacted in 1371, 1373, 1383, 1388, and 1389. The one exception was the statute of 1353, which promulgated the fees owed to the alnager and king for the alnager's services.

Officials and Enforcement

The almost systematic enactment and repeal of cloth measurement regulations between 1393 and 1411 made the task of the alnager extremely difficult. In the first fourteen years of this chaotic period merchants could sell cloths of any length and width as long as the "aulnage, subsidy, and other duties" were paid. Obviously, the alnager no longer had to measure cloth, but how was he supposed to determine the proper fees and subsidies? Since there is no provision concerning such matters in the act, one can only assume that the alnager was expected to use his own judgment. Perhaps he instituted an entirely new set of rates or found some way to apply those contained in the enactment of 1353. Furthermore, since nothing is said in the act concerning his other duties, London must have assumed that he would continue to inspect the quality of cloth in commercial transactions; to measure each piece or bolt with his copy of the standard ell, since this was the only way that he could assess properly the fees and duties; to mark and verify those judged to be of sufficient quality; to send those of insufficient quality to the king; and to retain a modest portion of the monetary proceeds to pay his staff. London was certainly leaving a vast array of important questions unanswered and displaying an inordinate amount of trust in one of its public servants.

An act in 1405 marked a partial return to cloth measurement regulations by reinstituting the 1328 regulations pertaining to white and colored cloths, but an act in 1407 argued that such restrictions were "trop grevous et damageous" and once again the alnager was instructed to ignore standard sizes for cloth. Notwithstanding this repeal, the act of 1405 was revived in 1409 and confirmed in 1411. During the latter legislative process the framers of the statutes used "aunes" and "verges" interchangeably, whereas one means "ells" and the other "yards." Such actions must have annoyed merchants and their customers and seriously weakened the effectiveness of the alnager. At best, cloth officials could attain exact knowledge of inexact laws. But even this must have been outside their reach since the repeals were not specific as to whether all cloth enactments were null and void and the reenactments referred generally only to one previous statute.

One year before the passage of the confirmation act of 1411 a new seal had been delivered to the alnager, but the document does not relate how it differed from the old one.[78] After "searching and surveying" cloth, the alnager was told to mark fabrics with the seal. Refusal to comply with these instructions would result in fines of £10 and £20 for the first and

---

78. The clause reads: "*Que* soit fait & delivere as Ulnerers suis ditz un novel Seal, eiant signe & merche different a la veile Seal de lour dit office" ["That a new seal be made and delivered to the aforesaid alnagers, having a different sign and mark from the old seal of their said office"] (*Rotuli parliamentorum,* 3:645).

second violations; third and subsequent violations merited imprisonment. Since the latter carried no time limitation such as one month or one year, the actual amount probably depended on the severity of the infraction and/or on the severity of the presiding judicial officer.[79]

Almost all cloth statutes after 1411 deal with measurement regulations, the most important being those of 1433, 1468, 1483, 1551, and 1708. Alnagers reached the height of their powers in the fifteenth century and lost most of their authority thereafter. They are mentioned in only five statutes after 1411. In two acts of 1439 and 1464 they are instructed to measure cloth with a cord that contained an additional inch to every yard. The same instructions are repeated in 1483, 1514, and 1549, except in these enactments Parliament advised that the inch be judged by the breadth of the alnager's thumb. Evidently it was during the 1550s that the alnager became a collector of cloth fees and duties. In 1557 an act of Queen Mary authorized searchers or sealers appointed under the act to enter houses in any corporate town to search for cloths. They were to determine whether fabrics were sealed with the alnager's and town's seals or whether they were deficient in length and breadth.

Source materials providing information on the day-to-day operations of alnagers are sparse indeed, but they are far more plentiful than what exists for port measurers and weighers. It is a commonly accepted fact that as early as 1300 London had a corps of master measurers of corn, supposedly eight in number, who with twenty-four assistants—three to each master—were responsible for insuring that buyers were not cheated by merchants in grain sales. Each master reportedly had certain standards by which merchants' measures were tested. The standards listed most frequently are the quarter, bushel, half-bushel, and strike. In addition, the capital employed somewhere between two and six master measurers of salt. Selected annually by the Salters' Company, they were supposed to provide standards and to regulate sales of salt by examining measures used by merchants. At approximately the same time London installed a

---

79. Apparently even these regulations and those confirmed in 1411 could not prevent foreign alnagers from impeding the alnage process as the following illustration from 1432 shows: "Please hit unto yowe the full wyse Communes of this present Parlement to considere, howe yat there as it is ordeignid and establid be diverses Statutes, that the Auneour shall mete, and enseale clothis or they be solde or putte to sale, with the Kynges seale there to ordeigned, up' peyne of forfaiture of theym. . . . Whiche not withstandyng, affir such cloth is mesurid, and enseled be the Auneour with the Kynges seal, in diverses Countres, comen the Auneours of other places, and forfete certeyn clothes so before mesurid, and ensealid . . . and the subside there of deuly payd, serchyng ageyn the lengthe and brede; and in yis wyse grevousely vexen the Marchauntes, and the Kynges people from day to day" (*Rotuli parliamentorum ut et petitiones, et placita in parliamento tempore Henrici R. V*, ed. Rev. John Strachey et al., 4 [London, 1832]: 404.

public weighing machine and appointed an official weighmaster to supervise operations.[80]

It is regretful that nothing is ever said concerning the actual operations. One does not know which ports had weighers and measurers. It is unclear whether these officials were supplied with standards from the Exchequer, from some other state depository, or from local guilds. Since it does not seem possible that such a small number of public servants could have supervised each and every sale, it would be helpful to know which ones were intended and whether volume or price was the more important ingredient. Granted that some types of operations were conducted periodically or throughout the entire year, one cannot be sure if the object of such operations was to insure compliance with national standardization programs or to enable the government to have a more accurate assessment of the volume of trade for taxation purposes. Perhaps the second was paramount; perhaps both were implied. Finally, none of the secondary sources mention whether these port authorities verified legal weights and measures; destroyed defective or fraudulent ones; or fined, punished, or incarcerated offenders.

A thorough examination of the primary and original sources does little to remedy most of these problems. In the earliest mention of port officials—found in the Patent Rolls and dating from 1339—one Nicholas de Ellerker was appointed as the weigher of wools for the port at Caernarvon. The entry merely states that Nicholas was to be paid "such fees as the king's weighers [*tronatores*] in the ports of England have used to receive."[81] In 1350 a Close Roll complained that Edward III had been defrauded "of his custom" by false weighing practices and that no punishment hitherto had been awarded. Consequently, Parliament demanded that "if the weigher be found to have made fraud in the weight of the wool and is attainted upon the deed, he shall have judgment of life and members. And . . . the merchant who has given anything to the weigher to make such fraud and is attainted for this, shall incur forfeiture of his wool so weighed."[82]

No mention of port measurers occurs before 1353. In a statute of that year, Parliament ruled that measures for red and white wine imported into England, Wales, and Ireland must be "lawfully" gauged by the king's measurers or their deputies. If these officials failed to perform any of their

---

80. See especially Salzman, *English Trade*, p. 56; W. Avery and T. Avery, *Suggestions for the Amendment of the Law Relating to Weights and Measures* (London, 1888), p. 8; and A. R. Bridbury, *England and the Salt Trade in the Later Middle Ages* (Oxford, 1955), p. 139.
81. *Calendar of the Patent Rolls: Edward III: 1338-1340* (London, 1898), p. 322.
82. *Calendar of the Close Rolls: Edward III: 1349-1354* (London, 1906), p. 222.

duties or performed them fraudulently, injured parties collected triple damages and offenders lost their assignments, were imprisoned, and could be ransomed only at the king's will. Still, what duties were expected of them? This illustration only affords proof of dereliction of duties. The same is true for a statute of 1357. Here Parliament attests again to the defaults of port measurers and reinforces the enactment of 1353 by threatening to forfeit the sellers' wines to the king if the sellers acquiesced knowingly to illegal practices. All of these provisions are repeated in a statute of 1380 and vinegar, oil, honey, and liquor measures are included.

No further mention of these officials can be found until the sixteenth century. Included in an Ordinance Book of 1565 is a list of rules for every port weigher to observe. For his salary of one mark per annum, the weigher was to be available at all "convenient and requisite" times to serve the needs of merchants and was personally responsible for the maintenance of the beam and scales. Repairs were to be authorized promptly by him. He was to supervise all weighings and could not lend his personal standards to anyone. Immediately after each weighing was completed, the merchandise (only wool is mentioned) had to be taken from the weighhouse and put to sale. Finally, all weighers had to take the following oath upon commencement of their appointments:

> Ye swere etc. Justiciable and obeisaunt unto the Maior his lieutenaunt Constables and companie of merchauntes of the staple in alle matters apperteyninge to the same estaple[.] Ready and attendaunt at all tymes requisite in the wayehowse to serve the merchaunt and to geve true and iuste weight according to thorders of the staple in sort that neither the merchaunt seller nor buyer maye be indammaged thoroughe your default[.] And alle other thinges to your saied office of wayer belonginge welle and truely you shalle do and perfourme to your power not letting for love or hatered winninge or losing or any other thinge that may be with you or against you so helpe you God.[83]

After 1565 the only reference to port weighers occurs in a source published in 1677: "There is in the most eminent *Cities* a publick Weighhouse set up and appointed, where every man may repair unto, either for necessity of *weighing,* or tryal of his *weights.* . . . The *Master, Overseer* or *Weigher* being ever sworn and deputed to do justice and right in his *weighing.*"[84] In the same source is this rather cryptic notice of the measurer: "There is for the reiglement of things *measurable,* instituted a publick *Measurer,* Authorized by the Sovereign Magistrate, who is sworn to decide all Controversies that happen in and about the Art of *measuring;*

83. E. E. Rich, *The Ordinance Book of the Merchants of the Staple* (Cambridge, 1937), pp. 119-20.
84. Roberts, *The Merchants Map,* p. 33.

to whose honesty and faith is intrusted this publick *measure,* and to which all *merchants* and *Traders* may in time of need and difference repair and have recourse unto."[85]

Before the organization of the inspectorate, the most influential weights and measures officials were clerks of the market. Representing the Crown in all aspects of metrological law for more than 400 years, they had no other functions and operated outside the jurisdiction of local political and judicial personnel. The clerk of the market for the king's household (*clericus mercate hospitii regis*) supervised the king's standards and was supposed to insure that local weights and measures conformed to them. Originally a deputy of the knight marshal, in the course of time he came to be regarded as the king's supervisor of markets, who, with a staff of assistants, regulated the weights and measures employed in commercial transactions.[86] There were clerks assigned to many counties as well. Chosen from among mayors, urban residents, and members of the feudal order, the local clerks appear to have been a later development resulting from the overflow of responsibilities placed on the king's clerk and from the hostility shown the king's clerk by irate local citizens.

The activities of clerks of the market are richly documented in the statutes and in letters patent and close. In the earliest surviving account, Edmund Biroun, clerk of the market for Edward I during the last decade of the thirteenth century, was sued by more than a dozen Irish towns for charging exorbitant fees for his services and for amercing innocent parties unjustly. When Edmund learned of this suit, he fled Ireland. The justiciar promptly ordered his arrest. After his capture in Wales, he was brought to Dublin, tried, and jailed in the castle of Dublin. Within a short time, however, Edmund escaped and Irish sheriffs were instructed to conduct a nationwide manhunt for him and to retry him upon his capture.[87] Later records do not indicate whether he was ever brought to justice.

The next reference to a clerk of the market occurs in 1321, when the city officials of London were ordered to show the king's clerk "the measures called 'the standards of London,'" and to permit him to assay them against the royal wine, ale, and corn measures. The unnamed clerk was supposed to determine which of London's measures were fraudulent and to order their replacement. The letter stipulated that immediate action was necessary because of the frequent appeals to the Crown from buyers who claimed that they were being cheated in the city's markets.[88]

85. Ibid., pp. 35-36.
86. For a discussion of the connection between this clerk and the office of knight marshal, see *Select Cases concerning the Law Merchant: A.D. 1239-1633*, pp. xlii-li.
87. *Calendar of the Justiciary Rolls*, 1:316.
88. *Calendar of the Close Rolls: Edward II: 1318-1323* (London, 1895), p. 362.

Almost the same situation is described in a document of 1338 pertaining to North Wales. Here irregularities were found in the measures for corn, wine, cheese, and certain other products. Edward III, on the advice of his council, ordained that weights and measures in agreement with the standards of England must be made and employed in all Welsh mercantile sales and purchases. To guarantee that these orders were implemented, he appointed one Richard de Kymberle to the office of clerk of the market for North Wales. Nothing is said concerning the scope of Richard's powers. His fees were left to the discretion of Edward's council.[89]

In the statute of 1340 that established commissions of private citizens throughout England to survey weights and measures, one clause protected the metrological powers of the clerks against local usurpation. This first statutory recognition of their authority, however, adds absolutely nothing to one's knowledge of their specific functions. Their rights were preserved from encroachment, but what were those rights? Since they were not described either in this enactment or in previous legislation, how were local commissioners and clerks supposed to determine at what point their jurisdictions overlapped? If the government never expected them to work in isolation, were they to coordinate certain of their functions and operate as a team? By saying nothing Parliament was stoking the fires of discord, and intrashire rivalries were instrumental in retarding London's well-intentioned standardization programs.

The singular dearth of such critically important data in the statute is in remarkable contrast to the wealth of information contained in the Patent and Fine Rolls for 1341-42. For instance, in the appointments of John de Ampleford as the king's clerk of the market and of Henry de Wivyll and Humphrey de Stowe as commissioners in Northamptonshire, all three men were ordered to examine the county's weights and measures and to hear and determine offenses committed. The same order is found in Ampleford's and other commissioners' appointments to the counties of York, Lincoln, Gloucester, Nottingham, Derby, Leicester, and Warwick.[90] None of the documents states specifically that operations were to be coordinated within each shire and later entries do not settle this problem. The king's clerk and the commissioners are merely addressed as a group.

That they actually operated as a group might be inferred from an order given Ampleford several months after the initial appointments. A letter commands the York delegation

> to survey that the measures and weights used in buying and selling . . . agree with the king's standard and to punish all convicted of false measures and

---

89. *Calendar of the Patent Rolls: Edward III: 1338-1340*, p. 322.

90. *Calendar of the Patent Rolls: Edward III: 1340-1343*, p. 310. The appointments for Somersetshire, Dorsetshire, and Lancastershire are on p. 446; they were made in 1342.

weights according to the statutes thereon. . . . To cause to be levied all fines, ransoms, issues forfeit and amercements taken and adjudged before them, as soon as they be brought to the wardrobe by the . . . clerk or his deputy and delivered to the king's clerk, William de Cusance, keeper of the wardrobe, by indenture. . . . The king will cause them to have due allowance in their account.[91]

This could be interpreted to mean that the clerk, possibly a supervisor of this particular commission, was responsible for coordinating the group's work schedule and, at the completion of each inspection, was to insure against fraud by delivering the group's proceeds directly to London. To fulfill the latter function it would have been advantageous for the clerk to be present at each session; otherwise he would have no certainty that he was delivering to Cusance every fine collected. The three officials may have inspected on some prearranged circuit, and the clerk was the government's assurance that every fine was turned in.

But this interpretation could be disputed in light of information supplied by a later entry in these accounts. Early in 1342 the government complained that one of its recently appointed commissioners, Thomas de Shirburn, had collected large fines in Lincolnshire and was "preparing to go to foreign parts with the said money and so defraud the king." In this instance, Ampleford and two others were empowered to arrest Shirburn wherever he was found and to imprison him until he made amends to the Exchequer for the money spent.[92] Here it would appear that the group actually did operate separately because Shirburn, himself, collected fines and, when the order for his arrest was issued, Ampleford obviously did not know of his immediate whereabouts.

Additional information on the development of the clerks' powers during these Edwardian years can be gleaned from the Rolls of Parliament. In an entry dated 1344, Commons limited the territory (called the verge) over which the king's clerk could determine pleas of trespass to 12 leagues in and around the site of his court.[93] Although the reasons for this action are never stated, three might have been of consequence. First, the limitation prohibits the clerk and his assistants from stationing themselves permanently in one area, thereby neglecting to perform their functions over the entire territory assigned them. Second, the number of cases at each court session is kept at a relatively small number, thus minimizing delays in stopovers along the clerk's circuit and allowing greater coverage of infractions within the territory during any particular year. Third, a deterrent to crime is provided since merchants and others could normally

91. *Calendar of the Fine Rolls: Edward III: 1337-1347* (London, 1915), p. 232.
92. *Calendar of the Patent Rolls: Edward III: 1340-1343*, p. 553.
93. *Rotuli parliamentorum*, 2:149.

expect a visit from this peripatetic band of officials at approximately the same time every year. In 1376 Commons repeated these instructions and insisted that clerks search out false weights and measures and pursue within the verge those whom they found in default. They were forbidden from taking excessive fines and their punishments had to be in proportion to the infractions committed.[94] Neither in these documents nor in any other, however, was there any indication of what constituted proper fines.

Statutory supervision of the activities of the king's clerk began formally in 1389. Chapter 4 of an enactment passed in the thirteenth year of Richard II's reign gave the clerk authority to burn false weights and measures. He could no longer take "common fines," but had to punish transgressors according to the severity of their infractions. He was not permitted to ride with more than six horses, and was not to delay in traveling from one town to another. The stipulation concerning the number of horses was based, perhaps, on Parliament's fear that a greater number would cause an unnecessary expense. It might also have been a roundabout way of limiting the number of assistants employed by the clerk. If the clerk disobeyed any of these instructions, he paid the king 100s. for the first offense, £10 for the second, and £20 for the third.

Three years later Parliament made it mandatory for the clerk to have brass weights and measures fashioned after the Exchequer standards and to have them signed and marked in conformance with Exchequer regulations. These were to be brought to the site of each local examination and no assay was legal without them. The statute never says whether the clerk or the government paid for these standards.

The statute of 1392 was the last parliamentary enactment to deal with clerks of the market until the seventeenth century. During the interim, state and local sources demonstrate unmistakably that the clerks became the most powerful metrological officials in British history. In 1399, for instance, John Foliambe was made crier in the court of the marshalsea at the same time that he was appointed the king's clerk of the market.[95] Twenty-five years later a knight, John de Pilkyngton, received in one grant the offices of escheator, clerk of the market, and keeper of the weights and measures.[96] All three of these positions pertained to Ireland. The document mentioned that another knight, Ralph de Standyssh, had held these offices before Pilkyngton. An account from 1451 described how the clerk examined the city's weights and measures, verified them against his own standards, and sealed those passing the inspection "withe a synet that longud to his offys."[97] Many other examples could be cited.

94. Ibid., pp. 336, 349.
95. *Calendar of the Patent Rolls: Henry IV: 1399-1401* (London, 1903), p. 80.
96. *Calendar of the Patent Rolls: Henry VI: 1422-1429* (Norwich, 1901), p. 51.
97. *Coventry Leet Book*, p. 267.

But if the sources reveal greater authority for the clerks by the middle of the fifteenth century, thereafter they testify almost solely to the misuse of that authority. In petitions to Parliament and government ministers, clerks are accused of violating most of the statutory injunctions, of overcharging on fees, of employing defective standards, of inflating fines, of harassing citizens and public magistrates, of extorting and blackmailing those punished for second or third violations, and of punishing summarily guilty and innocent alike. The abuses became so common that many mayors, aldermen, and burgesses took over urban metrological functions and barred the clerks from further service in their towns. So serious had these abuses become that a writer in 1635 pleaded that "one Justice of the peace at least ought to set with the Clarke of the market, to see that the Kings subjects bee not wronged."[98] Again, these examples could be multiplied many times over.

Such conditions forced Parliament to take corrective action. In chapter 19 of a statute of 1640 the king's clerk was confined to operations only within the verge of his majesty's court.[99] To alleviate local resentment, Parliament permitted mayors, bailiffs, lords of liberty, or their deputies or agents to exercise all of the functions of local clerks and to prohibit the king's clerk from entering their territorial jurisdictions. If these urban or feudal magistrates accepted and later defaulted, they were subject to fines of £5 for each infraction. If they received fines or fees, other than those permitted by statutory law, for inspecting and verifying weights and measures that formerly had been marked or sealed, they were fined £5 for the first offense, £10 for the second, and £20 thereafter. The same punishment applied anytime they imposed fines or amercements on merchants and producers without proper trials. Hence, they were not allowed to accept fees for merely examining weights and measures or for reexamining those that already had been verified and sealed.

98. Michael Dalton, *The Countrey Justice* (London, 1635), p. 146.
99. The opening paragraph of this act recites what Parliament considered to be the major abuses: "The undue execution of the Office of Clerk of the Market hath been very grievous unto divers of his Majesties most loving Subjects, who have been very much troubled by unnecessary Summons, and charged with exactions of divers sums of Money. . . . The said evils have partly arisen by means of an inequality of Weights and Measures throughout this Kingdome, and by granting and letting to Ferme the said Office of Clerk of the Market, and the Execution thereof in and through all or most of the several Counties . . . for great sums of Money. . . . The said Fermours or Grantees, by their unjust and undue proceedings in the said Office, do extort from his Majesties Subjects again, to their great impoverishment, and yet little or no redress at all in their said Weights and Measures, or any benefit thereby accruing to his Majesty." Shortly before this act, Richard Carew in the *Survey of Cornwall* (1811 ed), p. 220, wrote: "The clerk of the market's office hath been heretofore so abused by his deputies, to their private gain, that the same is tainted with a kind of discredit, which, notwithstanding, being rightly and duly executed, would work a reformation of many disorders, and a great good to the commonwealth."

This enactment placed the maintenance of local weights and measures directly into the hands of local officials and it amounted to a virtual repudiation of the clerks and their assistants. Without perhaps realizing it, Parliament was resurrecting a system of inspection and verification reminiscent of that which dominated the fourteenth century.

Legislation dealing with clerks of the market ended with the statute of 1640. Thereafter, their authority slowly waned until most, if not all, of their duties were taken from them by the close of the eighteenth century. This situation is evident as early as 1665, when one observer described it in these terms: "[their] Power be much lessened by the distribution of it to, and exercise of it by, Justices of Assize, of Oyer and Terminer, and Justices of Peace."[100] The king's clerk still had his court and he still sent warrants to sheriffs and bailiffs to empanel juries. But documents pay him little heed after 1665, and later he is never mentioned at all, except for a 1778 reference in the *Encyclopédie*.[101]

100. W. Sheppard, *Of the Office of the Clerk of the Market, of Weights and Measures, and of the Laws of Provision for Man and Beast* (London, 1665), p. 118.

101. *Encyclopédie ou dictionnaire raisonné des sciences, des arts et des métiers, par une société de gens de lettres,* ed. Denis Diderot (Geneva, 1778), 26:441.

3

The Tudor Era

Changing Conditions

The Late Middle Ages was a time of monumental change throughout the European world. During the fourteenth century the Great Famine of 1315-17 and the recurrent outbreaks of the Black Death after the early 1340s had the devastating effect of eliminating approximately a third of Europe's 75,000,000 people. England alone lost slightly more than 1,000,000 of her 3,000,000 inhabitants. Mercenary armies during the Hundred Years War (1337-1453) helped to cripple the French economy by destroying a considerable portion of the country's agricultural land and produce, while across the Channel overly zealous English monarchs poured enormous sums of money, which should have been allocated to domestic needs, into a war machine bent on retaining centuries-old Norman and Angevin landholdings. The Holy Roman Empire had collapsed into its traditional feuding duchies after the demise of the House of Hohenstaufen in 1254, and Switzerland, its most valuable possession, was now at war striving for political and economic independence. Huge Mongol and Turkish armies had virtually shut off the East to Western merchants, particularly those from northern Italy, whose ancestors two centuries before had created the Commercial Revolution. Changes in warfare that included the introduction of the longbow, the pike, and gunpowder sounded the knell for the medieval knight and for the feudal political system. The emergence of cash-crop agriculture over vast stretches of Europe, the steady proliferation of towns, the growth of larger industrial complexes, the increased use of credit and of gold currency, and other important economic developments helped to eradicate the stranglehold of manorialism over Europe's rural life. For most of the fourteenth century the administration of the church was located in Avignon, France (the Babylonian Captivity, 1309-77); and that period was followed by more than a quarter century of schism, intense national rivalries, intermittent war, and general religious stagnation. In almost every facet of life change was overtly manifesting itself and nowhere was that phenomenon more pronounced than in England.

Most historians agree that the fifteenth century was politically and economically disastrous for England. Having lost all of its French possessions except Calais by 1450, it entered, after the war's termination, into three decades of fierce civil strife antedating the succession of the Tudors in 1485. Wool exports declined drastically due to such chaotic internal and external conditions, while cloth exports struggled with foreign competition. During the Wars of the Roses the Hanseatic League took advantage of England's domestic crises to reinforce its monopoly over regional and interregional trade in the northern markets. With the closing of many of these commercial centers to English traders, agricultural profits fell and landlords were forced into experimentation with leaseholds.

Disadvantageous as some of these conditions were, however, England did experience notable gains during the period. For instance, even though the war with France required large amounts of capital procured through additional taxes and foreign loans, the damage, exclusive of manpower losses, was sustained only by the French since the war was fought on French soil. Freedom from invasion during these two centuries helped English industry and town life to develop and made England so politically and geographically insular that it was necessary later to seek expansion and colonization overseas. In order to obtain funds for the French campaigns, English monarchs had to accede time and again to Parliament's demand for greater power over the government's purse-strings. Parliament, heavily staffed by mercantile and industrial elements, only awarded funds for continued war when the control of taxation became its responsibility. England was the only European country to award such powers to merchants and other businessmen during this period and a precedent was set that would have important ramifications under the Tudors. In other sectors, poor land that yielded little in agricultural produce was gradually turned over to sheep raising, and during the sixteenth century this industry would become an important source of national wealth. Forest reserves became sufficiently depleted during the later Middle Ages to force a greater reliance upon coal as a source of fuel and energy. The use of coal led to larger production units and helped entrepreneurs, because of the high cost of coal-using equipment such as the blast furnace, to make larger investments in industrial undertakings. With an area of 50,000 square miles—one-fourth as large as France but four times that of Holland—England would find that it possessed the necessary requirements for rapid economic advancement by the middle of the sixteenth century.

Under the Tudors (1485–1603) rapid economic gains were made in commerce, agriculture, and industry. All three of these sectors prospered because exploration and settlement of overseas areas produced vast new markets and gave England supplies of goods that helped create many new industries. English landlords saw the financial benefits to be derived from their estates: they increased their arable acreages through the clearing of wastes and deforestation, and they exploited systematically their deposits of iron ore, coal, and stone. Both squires and yeomen became agricultural financiers. In the towns, industries diversified and took on new machinery and new workers. Textile manufacture proliferated in farmsteads and villages where the cost of living was low, where spinners were abundant, where guild regulations were absent, and where there were excellent sources of water power. In Lancashire alone there arose an elaborate network of middlemen who supplied the basic materials to thousands of domestics who labored at spinning wheels and looms. There was a substantial increase in iron production due to the introduction of blast furnaces and water-wheel-operated bellows. Gun casting became a valuable English enterprise. Sheffield emerged as a European center for cutlery, and Birmingham for such small parts as knives, locks, and buckles. Appearing for the first time were paper and gunpowder mills, cannon foundries, alum and copperas factories, sugar refineries, and large-scale saltpeter works. The production of zinc, iron, copper, lead, and tin rose significantly. The discovery of calamine, the ore of zinc, in Somerset and elsewhere, together with the first really effective attempts to mine copper ore, made possible the establishment of brass-making and battery works for hammering brass and copper ingots into plates. The latter developments led eventually to standardized sizes for ingots, sheets, rods, and wire. Cutting mills were built for producing iron rods for nailers, smiths, and shipwrights, and water-driven machinery was adopted for hammering metal bars in various forms.

When the gold and silver bullion imported from the New World found its way into European mints a price inflation set in, which helped to stimulate the growth of these various industries. Since the supplies of gold and silver increased more rapidly than the existing stock of goods, the rate at which money circulated was accelerated and many people, fearing shortages and higher prices, used their available cash for immediate rather than future purchases, thus stimulating the economy. In just one industry, the price of wool rose so rapidly that the prospect of profits, coupled with the shortage of labor and low wages, induced landowners to specialize entirely in sheep raising. Hence a price inflation produced a corresponding profit inflation. As the center of commerce shifted from the Mediterranean to the Atlantic seaboard, and as colonialism led to fundamental changes in business organization, since large-scale overseas enterprises required greater sums than could be provided by the old associations, a new form of business enterprise came into being—the stock or joint-stock company. The Merchant Adventurers company—reorganized but dating back to the later Middle Ages—shipped cloth to markets between the mouth of the Somme, in northeastern France, and the tip of Denmark. The Eastland Company (1579) monopolized trade with Scandinavia, Poland, and the eastern coast of the Baltic, and the Muscovy (1555) and Levant (1581) companies cornered the Russian and Turkish markets respectively. In 1600 the largest of these merchants' associations—the East India Company—was founded. English merchants, like their medieval Italian counterparts, bought and sold on commission for their continental patrons. They shipped goods or imported them through continental agents. They sent their own agents to all parts of the world. New products arriving in English ports included such items as tea and opium from the East, the guinea hen from Africa, and cocoa, sugar, tobacco, potatoes, codfish, maize, and tomatoes from the New World. In the process London became the most important city in Europe and served simultaneously as a capital, metropolis, port, and entrepôt. Its population increased from slightly less than 50,000 in 1500 to well over 200,000 by 1600.

With all of these developments occurring in the economic and social life of the English during the sixteenth century, it is not surprising that a new metrological era was ushered in. Earlier governments had issued an immense number of enactments whose principal purpose was the establishment of a uniform and well-regulated system of weights and measures. Uniformity meant cohesiveness among the many units inherited from other societies, while regulation signified the expansion of metrological administrative duties among many diverse groups and professions. The Tudor governments would continue to issue laws regulating those units, although at a much slower pace, and they would contribute to the further growth of the administrative machine. But their fame was not derived from their contributions in these areas. It was achieved almost solely from the construction, design, function, and dissemination of physical standards.

Early Tudor Standards

Before the imperial weights and measures era began in the third decade of the nineteenth century, no period in English history was as important from the standpoint of physical standards as the Tudor. Emanating principally from the parliamentary legislation and orders-in-council of Henry VII (1485–1509) and Elizabeth I (1558–1603), Tudor standards were among the most precise and sophisticated in Europe. Built generally of brass or bronze, they were duplicated and distributed to all parts of the kingdom. Unfortunately, they were accompanied all too often by instructions specifying that the old weights and measures were to be broken up and destroyed. These periodic demolitions, together with the well-established practice of melting down old standards every time new ones were requested, are chiefly to blame for the sparse numbers of medieval and early modern weights and measures now available.

The history of the Tudor standards commenced in 1491. In this seventh year of Henry VII's reign, Parliament declared that past attempts had failed to accomplish metrological uniformity and that only a massive assault on local customs could eradicate the innumerable variations and the resultant fraudulent weighing and measuring practices.[1] Desiring to place all weights and measures on a clearly defined statutory basis, Parliament ordered the construction of new Exchequer standards of weight, length, and capacity. Stored in the Treasury and supervised by the commissioners of assize, they became the Tudor primary reference standards. By a statute passed in 1495, copies of them were supplied to 43 shire towns in England.[2] The chief officers of these metropolitan centers

1. For examples of the many types of fraudulent practices and of illegal weights and measures, see *Calendar of the Close Rolls: Richard II: 1381-1385* (London, 1920), p. 365; ibid., *1392-1396* (London, 1925), p. 424; *Calendar of the Patent Rolls: Edward III: 1345-1348* (London, 1903), p. 535; Hubert Hall and Frieda J. Nicholas, eds., *Select Tracts and Table Books Relating to English Weights and Measures (1100-1742)*, Camden Third Series, 41 (London, 1929): 10; *Munimenta gildhallæ Londoniensis: Liber Albus, Liber Custumarum et Liber Horn,* ed. Henry Thomas Riley (London, 1859-62), p. 348; *Memorials of London and London Life in the XIIIth, XIVth and XVth Centuries,* ed. Henry Thomas Riley (London, 1868), p. 121; *Medieval Archives of the University of Oxford,* ed. Rev. H. E. Salter (Oxford, 1921), p. 267; *Monumenta Juridica: The Black Book of the Admiralty,* ed. Sir Travers Twiss (London, 1873), pp. 80, 163; "Palatinate of Chester: 1351-1365," in Part 3 of *Register of Edward the Black Prince* (London, 1932), p. 28; *Rotuli parliamentorum ut et petitiones, et placita in parliamento tempore, Edwardi R. I,* ed. Rev. John Strachey et al., 1 (London, 1832): 47; *Select Pleas in Manorial and Other Seignorial Courts,* ed. F. W. Maitland, Selden Society Publication, 28 (London, 1889): 11; *Select Cases concerning the Law Merchant: A.D. 1270-1638,* ed. Charles Gross, Selden Society Publication, 23 (London, 1908): 19, 40, 62.

2. The towns designated were Appleby, Newcastle, Carlisle, Lancaster, York, Lincoln, Derby, Nottingham, Leicester, Coventry, Uppingham, Northampton, Bedford, Buckingham, Cambridge, Huntingdon, Norwich, Bury St. Edmunds, Chelmsford, Hertford, Westminster, Maidstone, Guildford, Lewes, Oxford, Reading, Shrewsbury, Stafford, Hereford, Gloucester, Worcester, Salisbury, Winchester, Ilchester, Dorchester, Exeter, Lostwithiel, London, Bristol, Castle of Dover, Southampton, and Chester. Other cities, however, received copies of these standards, including the following in Ireland: Dublin, Drogheda, Carlow (Catherlagh), Dundalk (Dundalke), Daingean (Phillipstowne), Trim (Tryme), Kildare, Wexford, Mullingar (Molingare), and Port Laois (Maryborough). The

Figure 1. English linear measures from 1497 to 1844. Top to bottom: latten metal yard (1497) of Henry VII; Exchequer yard and ell bed (1588) of Elizabeth I; Exchequer yard (1588) of Elizabeth I; Exchequer ell (1588) of Elizabeth I; yard (1659) of the Tower of London; Graham's Royal Society yard of 1742; General Roy's 42-inch scale of 1784; Captain Kater's imperial yard bed of 1824; Captain Kater's imperial line yard of 1824; Baily's imperial bronze line yard of 1844. (Crown Copyright. Science Museum, London)

had to pay for them, assume responsibility for their safekeeping, and supervise the repairing or remaking of local standards. All assized weights and measures were stamped with special seals in the presence of justices of the peace, who served as overseers for the central government to insure that local authorities complied with these regulations.

The program failed. In 1496 Parliament issued a new statute declaring that the standards sent to the shires were defective and had to be returned. They were melted down and new ones were constructed in 1497. Over the course of the next two years Parliament redistributed them and once again expenses were paid by the chief magistrates of the towns.

The standard yard of this 1497 manufacture is preserved at the Science Museum in London (see Figure 1). Modeled after the standard of Edward I, it is a well-worn, octagonal, brass rod, half an inch thick,

---

weights and measures of England were established by law in Ireland as early as 1351, and one year before the 1495 enactment the statutes of England relating to weights and measures were made applicable in Ireland. See Richard Bolton, *A Iustice of Peace for Ireland, Consisting of Two Bookes* (Dublin, 1638), p. 279; John Quincy Adams, *Report of the Secretary of State upon Weights and Measures Prepared in Obedience to a Resolution of the Fourteenth of December, 1819* (Washington, D.C., 1821), H. R. Document 109, p. 27; David Macpherson, *Annals of Commerce, Manufactures, Fisheries and Navigation*, 4 vols. (London, 1805), 1:545; and "Seventh Annual Report of the Warden of the Standards on the Proceedings and Business of the Standard Weights and Measures Department of the Board of Trade for 1872-73," in *Parliamentary Papers*, Great Britain, *Reports from Commissioners*, 38 (London, 1873): 48.

Early Tudor Standards

measuring one yard from end to end, and divided into three one-foot divisions. One of the feet is marked in 12 inches and the entire yard is subdivided by rather coarse lines into $\frac{1}{16}$, $\frac{1}{8}$, $\frac{1}{4}$, and $\frac{1}{2}$ of a yard. The overall length of the rod is only .037 inch shorter than the current imperial standard. Close to each end is a boldly stamped Gothic letter *H*. This standard was constantly in use for testing by the government until a new yard standard was made in 1588.[3]

The statute of 1496, however, was concerned chiefly with capacity measurement standards. It provided new definitions for the bushel and gallon of wheat, based upon the troy weight system.

1 bushel	=	8 gallons of wheat
1 gallon	=	8 troy pounds of wheat
1 troy pound	=	12 troy ounces
1 troy ounce	=	20 sterlings
1 sterling	=	32 grains of wheat taken from the middle of the ear

As in the *Tractatus* of Edward I, the wording is ambiguous and misleading. Relying on formulas found in other late-fifteenth-century documents, this statutory provision should read:

1 bushel of wheat	=	8 gallons of wheat
1 gallon of wheat	=	(a) a striked measure containing 8 troy pounds of wheat [replacing the formula of 8 tower pounds in the *Tractatus*], or (b) a vessel large enough to contain, when striked, a cubic capacity equivalent to the area of 8 troy pounds of wheat
1 troy pound	=	12 troy ounces
1 troy ounce	=	20 sterlings or pennyweights, each sterling being a term of account and reckoned as weighing 24 grains
1 sterling or pennyweight	=	[only for convenience' sake and to establish a link with the *Tractatus*] the approximate equivalent of 32 average-sized wheat grains, which equaled 22.5 troy grains or barleycorns

Or, these units can all be described quite simply in troy grains:

1 pennyweight	=	24 grains
1 ounce	=	480 grains (20 pennyweights of 24 grains each)
1 pound	=	5760 grains (12 ounces of 480 grains each)
1 gallon (in weight of content)	=	46,080 grains (100 ounces or 8 pounds)
1 bushel (in weight of content)	=	368,640 grains (800 ounces or 64 pounds)

3. H. W. Chisholm, in "Seventh Annual Report of the Warden of the Standards," p. 34, complained that "independently of the coarseness of construction . . . the process of stamping it with an H necessarily made a sensible alteration in its actual length."

Table 1
A Comparison of the Troy and Tower Weight Systems

Unit	Troy Weight	Tower Weight	Difference
1 pennyweight	24 grains	22.5 grains	1.5 grains
1 ounce	480 grains	450 grains	30 grains *or* 0.0625 troy ounce
1 pound	5,760 grains	5,400 grains	360 grains *or* 0.75 troy ounce
1 gallon (in weight of contents)	46,080 grains	43,200 grains	2,880 grains *or* 6 troy ounces *or* 0.5 troy pound
1 bushel (in weight of contents)	368,640 grains	345,600 grains	23,040 grains *or* 48 troy ounces *or* 4 troy pounds

The difference between the troy and the tower systems and their effect on the statutory definitions of the gallon and bushel are given in Table 1.

The 1496 enactment made the troy system a legal, but not the exclusive, English basis for weighing. The document merely described it and then used it only to establish unit standardizations for the grain gallon and bushel. The tower pound, ostensibly, was still considered by the government as the standard for money, precious stones, gold, silk, silver, and electuaries. All other items were weighed according to the mercantile system or the recently introduced avoirdupois. The mercantile system, employed by merchants since at least the late twelfth century, was an outgrowth of the tower pound because it consisted of 15 tower ounces of 450 troy grains each, or 6750 troy grains in all. This pound was slowly replaced during the sixteenth century by the avoirdupois pound of 7000 grains. No statute was ever issued that defined the avoirdupois system and prohibited use of the mercantile, but in the nineteenth year of Henry VIII's reign (1509-47), Parliament formally abolished the tower pound. This act of 1527 made the troy pound the legal standard for those items previously assigned tower weight.

Further problems remained, however. Following the abolition of the tower pound and the replacement of the mercantile by the avoirdupois system, two other commercial pounds had arisen. One of these, a pound of 7680 grains (16 troy ounces of 480 grains), had been created inadvertently by a statute of Henry VII in 1497 (troy ounces had been used instead of avoirdupois ounces). It was employed for weighing ordinary merchandise other than precious metals, bread, or wool. The other was the 7200-grain merchant's pound of the Hanseatic League, the double mark of Cologne and Hamburg, used at its London establishment, the

Steelyard. Equal to 16 tower ounces of 450 grains, or to 15 troy ounces of 480 grains, it had come to be regarded as the replacement for the old standard of 15 tower ounces.

Both of these newly introduced pounds were doomed to an early death, the first because the error would be eliminated in future enactments, the second because the position of the Hanseatic League had already begun to deteriorate by the late fifteenth century due principally to internal troubles in its organization. After its members secured a renewal of their special trading privileges in England in 1474, their autocratic position had been weakened progressively, to the advantage of English merchants, by the time of Henry VII. Finally Elizabeth withdrew all of their privileges, allowing them only to retain the Steelyard in London in return for which she forced them to concede an English "factory" in Hamburg. In the interim, the avoirdupois pound replaced both of the new competitors.

Figure 2. Exchequer bronze gallon and bushel (1497) of Henry VII. (Crown Copyright. Science Museum, London)

Of the capacity measures constructed in accordance with the provisions of the 1496 enactment, only a gallon and several bushels still survive; they are displayed at the London Science Museum (see Figure 2). They are the oldest British standard capacity measures in existence. Each is of heavy bronze. The gallon contains an elegant handle, perpendicular to the measure. Each bushel is fitted with two handles, rather roughly attached to the bowl, and rests on three short, stubby feet. The words "Henricus septimus" are embossed on the gallon and the title "Henricus septimus dei gracis rex Hanglie et Francie" encircles the bushels. All of the lettering is in Gothic script and interspersed are designs of the Winchester portcullis, the Tudor rose, and a greyhound. The gallon and one of the bushels were gauged in 1931 by the Standards Department of the Board of Trade and the two measures were found to be 268.43 cubic inches and 2144.8 cubic inches respectively.

No troy weights were constructed following the statute, but the Exchequer did issue a series of avoirdupois bell-shaped and flat bronze

Figure 3. Avoirdupois bell-shaped and flat bronze weights (1497) of Henry VII: (a) 28 pounds, (b) 14 pounds, (c) 112 pounds, (d) 14 pounds, (e) 7 pounds, (f) 7 pounds, (g) 56 pounds. (Photo. Science Museum, London. By Courtesy of University Museum of Archaeology and Ethnology, Cambridge)

weights of 7, 14, 28, 56, and 112 pounds (see Figure 3). Each of these delicately fashioned standards bears a prominent relief either of the royal coat of arms or of the Tudor rose. Crowns, greyhounds, and several other objects augment these basic designs. Firmly attached to the crest of every bell-shaped weight is an oversized, and somewhat awkward, ring. Very few of these weights have survived.[4]

Tudor Officials

The acts of Henry VII refer constantly to urban officials because it was during the Tudor era that they reached the zenith of their metrological authority. Having received most of their duties during the medieval period, they were entrusted by the Tudor governments with the additional task of carrying out all statutory directives. In the act of 1491, for example, the chief officers of cities, shire towns, and boroughs secured the safekeeping of the new standards recently sent to them by London. They were instructed further to "correct and reform" the older weights and measures of their districts. Four years later Parliament reinforced these instructions by asking the same officers to "view, examine, print, sign, and mark" all of their standards. At least twice a year they examined weights and measures within their jurisdictions and burned or otherwise destroyed defective ones. They were also authorized to levy fines for each and every successive infraction, but London established the rates. By 1512, if offenders were unable to muster up the full amount of the fines, town officers could incarcerate them. All of these duties were conferred on the chief officers of Irish urban centers by a statute passed in the twelfth year of Elizabeth's reign. Scotland followed suit thereafter but it was not until the passage of two acts, in 1607 and 1609, that Scottish urban officials exercised metrological authority equal to their Irish and English neighbors.

The position of these officials in weights and measures administration peaked in the sixteenth century, waned in the seventeenth, and all but disappeared in the eighteenth. It is not surprising that this sequence of events should have occurred. First, as cities expanded both demographically and territorially, the political, economic, and social responsibilities of elected officers became too complex and time-consuming to be diverted to other pursuits, regardless of the latter's importance. Their energies, by necessity, were thrust into solving much larger and more immediately vital problems than their medieval forebears had ever known. Second, the mounting industrial and technological revolutions, the drastic agricultural changes, and other early modern phenomena had truly profound effects on the entire structure of British metrology. To administer this compli-

---

4. Chisholm remarked that the Great Fire of 1666 may have been responsible in part for this dearth ("Seventh Annual Report of the Warden of the Standards," p. 31).

cated system, professional examiners, later called inspectors, were required. Weights and measures administration had become a full-time job, needing well-trained and expert practitioners. It no longer could be delegated to untrained and overworked, albeit conscientious, mayors, bailiffs, aldermen, and others. It had, in a sense, come of age and the age demanded a thorough and complete change in operations.

The inspectorate took over the powers of elected officials in the eighteenth century, but during the Tudor period the officials' only serious competitors were craft guildsmen. Because of the innumerable variations in the capacities of barrels, kilderkins, and firkins made by ale and beer brewers, Parliament in 1531 ordered that capacity measures for malted beverages be made only by members of cooperage guilds. In particular, the London guild had to search, examine, and gauge all such measures in the city of London and its suburbs. If its members found vessels that did not conform or could not be altered to conform to statutory specifications, they confiscated and burned them. Every cooper had to place his own distinct mark or seal upon each cask he constructed so that the maker and the place of origin could be easily identified by buyers, merchants, and alnage officials.

The coopers gained considerable authority by this act but the effectiveness of their administration was seriously impaired by two provisions. First, one chapter permitted them to construct casks whose dimensions varied from the specifications in an earlier chapter so long as the makers marked the total number of gallons somewhere on the vessels. Parliament thus promulgated a list of acceptable gallon capacities for ale and beer measures, but told coopers that they did not have to abide by it. Second, every measure was described in terms of the "king's standard gallon." This was never defined and one can only presume that the clause referred to some contemporary physical standard which unfortunately has not survived. The measures identified by the statute are:

Measure	Ale	Beer
firkin	8 gallons	9 gallons
kilderkin	16 gallons	18 gallons
barrel	32 gallons	36 gallons

This act also contained several chapters on soap measures. The coopers' functions prescribed earlier for ale and beer measures applied verbatim in this section. The units described are:

firkin	=	8 gallons (the empty cask to weigh 6½ pounds)
kilderkin	=	16 gallons (the empty cask to weigh 13 pounds)
barrel	=	32 gallons (the empty cask to weigh 26 pounds)

There are three additional problems with these provisions. Even though all soap measures traditionally conformed to the specifications for corresponding ale measures, the gallon referred to here was not defined, nor was it described as the equivalent of the "king's standard gallon." One can only assume that the latter applied. In every case the description of the vessel contains the words "or above" immediately following the number of gallons. If the barrel, for example, was fixed at 32 gallons or above, why did Parliament promulgate a weight standard for the empty cask? An empty vessel built to hold 40 gallons surely would be heavier than one designed for a 32-gallon capacity unless, and this was not likely, lighter materials were used to construct the larger vessel. Parliament obviously was establishing minimum and not maximum gallon capacities for soap measures, but the weights given for the casks always are followed by the words "and not above." Consequently, a 40-gallon soap barrel conformed to the statute since no limit was set on how large it could be, but the weight of the cask, rigidly confined by the act, became meaningless. Finally, nowhere in the statute is there a clue as to whether the casks are to be weighed according to the troy, mercantile, or avoirdupois pounds. When a parliamentary enactment contained such contradictory and ambiguous wording, it is little wonder that administrators aroused the wrath of the general public and that governmental expectations for nationwide uniformity never were realized.

Just as in the case of town officials, craft guildsmen had received many of their metrological duties during the later Middle Ages. Operating under royal charters, these guilds controlled the inspections and verifications of the weights and measures used in their day-to-day operations. Such grants were fairly common from the fourteenth through the seventeenth centuries, with the London guilds securing the greatest share of them. The loss of such privileges by the eighteenth century was due to the same causes that undermined the metrological authority of the town magistrates, and the guilds' powers, too, would eventually fall under the aegis of the inspectors.

The earliest mention of acquisition by a craft guild of weights and measures privileges occurred in 1327, during the reign of Edward III. Reinforced by an additional charter under Richard II in 1392, the Goldsmiths' Company supervised London's tower and troy weights until 1679. Besides the usual inspection and verification duties, this guild conducted the testing of new coinage of the realm at the annual trial of the Pyx, held, for a long time, at their hall on Foster Lane in London. They also assisted the royal commissions that established the Elizabethan weight standards of 1588.

By the time the London Coopers' Guild had been awarded the franchise of inspecting and verifying the capital city's ale, beer, and soap measures in 1531, the membership already had a distinguished reputation for maintaining and obeying strict regulations in the construction of casks. In 1396, for example, the coopers applied to the mayor and aldermen for an ordinance restraining certain members of the union from making illegal vessels for beer or other malted beverages out of oil or soap tuns. Slightly more than a decade later they made another application ordaining that no one in the guild should make vessels for liquor unless the wood used was of the finest quality. An ordinance issued in 1420 required each cooper to have his own sign, made of iron, to mark every vessel that he constructed, a copy of the sign being properly registered in the guildhall's records.[5] In yet another ordinance, the mayor and aldermen in 1457 ordered them to burn both defective casks and those constructed of poor or marginal materials.[6]

The statute that awarded the franchise to the coopers in 1531 stipulated that the wardens and one official from the mayor's office would have authority, whenever they thought it convenient and expedient, to "search, view & gage" any ale, beer, and soap measures in London, its suburbs, and within a two-mile circumference outside the metropolitan limits. They were to check specifically for proper workmanship, "good and seasonable" woods used in construction, legible markings, and accurate contents listings. Every vessel that conformed to statutory specifications would be marked with the sign of St. Anthony's Cross. They could take one farthing for each completed test and could withhold verification on any vessel until they were assured that it had been repaired to their satisfaction. If a vessel could not be altered sufficiently to conform to the standard, they were authorized to burn it and to collect 12d. fine from the owner. In addition, only authorized coopers were permitted to construct such casks in the future, thereby removing this traditional practice from the leading breweries. Any cooper who made vessels

---

5. In a collection of cooperage documents, published by the Coopers Company, London, and entitled *Historical Memoranda, Charters, Documents, and Extracts, from the Records of the Corporation and the Books of the Company, 1396-1848* (London, 1848), p. 11, there are illustrations of some of these signs.

6. Such regulations were common also to the Turners' Guild, as the following citation from the *Memorials of London,* p. 78, shows: "Henry the turner . . . Richard the turner, John the turner, . . . Robert the turner . . . in the future . . . will not make any other measures than gallons, potells, and quarts; and that they will make no false measures, such as . . . "chopyns" and "gylles"; nor will they make them in the shape of boxes or of cups. . . . And that all such false measures, of whatever kind they may be, and, wheresoever they may be found, whether in the hands of foreigners as well as of freemen, they will attach, and will cause the same to be brought to the Guildhall, before the Mayor, and present the same, on pain of heavy amercement."

contrary to this statute was liable to a fine of 3s. 4d. for each infraction, one half going to the king, the other half to any citizen who pressed charges.

In 1589 the company's jurisdiction was strengthened by the requirement that its members visit beerhouses within the territorial limits mentioned above to gauge and mark casks. They had 48 hours to comply with any such request made by the government or they were penalized 20s. No other enactments were made on their behalf after 1589 but their records indicate that they retained these privileges intact until the 1750s.[7] Scotland extended similar powers to one cooper in every borough by an act of 1693.[8]

Special provisions for the inspection of weights and balances of pewter or brass by the Pewterers' Company of London were outlined in two statutes of 1503 and 1512. In addition to those powers given the coopers, the master and wardens of this guild were permitted to appoint inspectors to fulfill the obligations of their charter. Lead weights came under the control of the Plumbers' Company a century later. In 1611, an act of James I authorized the company to "search, correct, reform, amend, assay, and try" all lead weights in London, its suburbs, and within seven miles of the metropolitan area. Its members could enter any house or business establishment to conduct their tests.[9] The company had an office at the guildhall and a member permanently stationed there to size and seal weights. He had custody of the scales used in the examinations. As late as 1757 there is evidence that such operations were still being conducted. With the discontinuance of the use of pewter and lead as materials for standards manufacture, the metrological privileges of both these guilds gradually disappeared by the end of the eighteenth century.

The last of the major London craft guilds to be given metrological duties was the Founders' Company. By a charter of 1614, it received permission to size and mark all avoirdupois weights sold or used in London, its suburbs, and within three miles of the metropolitan area. It had authority to search out defective brass weights and to destroy them. Until 1889 all brass weights sold or used in London were required to be

---

7. For a discussion of the reasons behind the guild's loss of these inspection and verification duties, see "Report from the Committee Appointed to Inquire into the Original Standards of Weights and Measures in this Kingdom, and to Consider the Laws Relating Thereto," in *Parliamentary Papers,* Great Britain, *Reports from Committees of the House of Commons,* 2 (1737-65): 427.

8. *The Acts of the Parliaments of Scotland: A.D. MDCLXXXIX-MDCXCV,* Great Britain Record Commission Publications (London, 1822), p. 260.

9. In the "Report from the Committee Appointed to Inquire into the Original Standards," p. 427, there is a list of fees that were customarily charged for assaying the various denominations of weights.

Figure 4. Exchequer avoirdupois bell-shaped and flat gunmetal weights, first series (1558) of Elizabeth I. The denominations are 1, 2, 4, 7, 14, and 28 pounds. (Photo. Science Museum, London)

stamped by its members in collaboration with an official stamper appointed by the Corporation of London.

The Elizabethan Standards

The most significant weights and measures developments of the Tudor era occurred during the reign of Elizabeth. Concentrating almost totally on the construction, inspection, and verification of physical standards, Elizabeth, through orders-in-council, impaneled special commissions to supervise the examination of weights and measures everywhere in the realm. The commissioners destroyed any standards that did not conform to Crown specifications and constructed new sets which they distributed to the Exchequer and to various local authorities.

The history of the Elizabethan standards began the very year Elizabeth succeeded Mary as queen of England. In 1558 the Exchequer received its first series of Elizabethan avoirdupois bell-shaped and flat weights in denominations of 1, 2, 4, 7, 14, 28, and 56 pounds (see Figure 4). They are

Table 2
Weights of Principal Standards Used in Various Locations in 1574

Weights of Standards in Avoirdupois Pounds					
Exchequer (Elizabeth I)	London (Henry VII)	Exeter (Henry VII)	Worcester (Henry VII)	Winchester (Edward III)	Norwich (Henry VII)
56	56	56	6	56	56
28	28	28	5	28	28
14	14	14	4	14	14
7	7	7		7	7
4	½	5			
2					
1					

Avoirdupois Weights Used by the Clerk of the Market for the King's Household		
Pounds	Ounces	
14	4	½
7	2	¼
	1	⅛

Troy Weights Used by the Clerk of the Market for the King's Household			
Pounds	Ounces		
None	48	8[a]	4[a]
	32[a]		2[a]
	16[a]		¼[a]

Avoirdupois Weights Deposited in the Tower of London[b]	
Pounds	Ounces
30	12
20	8
10	4
4	2

Troy Weights Deposited in the Tower of London[c]	
Pounds	Ounces
3	1
2	½
	¼
	⅛
	1 farthing

Sources: Sir Richard Glazebrook, "Standards of Measurement: Their History and Development," *Nature*, 128 (1931):18-21; and "Seventh Annual Report of the Warden of the Standards on the Proceedings and Business of the Standard Weights and Measures Department of the Board of Trade for 1872-73," in *Parliamentary Papers*, Great Britain: *Reports from Commissioners*, 38 (1873): xii-xiv, 11-15, 18-26, 27-32.

[a]Marked with a Gothic, crowned *M*.

[b]Also deposited at the Tower was a set of four weights (30, 20, 16, and 15 pounds) that was no longer used.

[c]Also deposited at the Tower was a set of eight weights (12, 10, 5, 5, 4, 3, 2, and 1 pounds) that was no longer used.

crudely made of gunmetal, with little of the expertise in emblem design and symmetry characteristic of the standards of Henry VII. Each of them contains a crowned Gothic letter *E* and a crowned Tudor rose. The rings on the bell-shaped measures, however, are more proportional to their weights than those on the set constructed in 1497.[10] There is no documentary evidence to show whether copies were made of these particular standards and distributed elsewhere. The weights mentioned in the proceedings of the jury of 1574, even though they were collected from only six locations, would tend to prove otherwise, since the Exchequer alone is singled out as having Elizabethan standards. Of the other five locations, four had weights dating from Henry VII's time and one had weights stamped "Edward III."

In 1574 Crown and Parliament realized that these divergent sets of standards were causing confusion. To remedy this unnecessary and costly situation, Elizabeth impaneled a jury of nine merchants and twelve goldsmiths—all London residents. The commission was entrusted with determining how many different types of weights were used in England; what their construction materials consisted of; what systems—if any—they were based on; where they were located; and what their names were. The commissioners found many discrepancies between the Exchequer weights and the standards of the principal cities of London, Exeter, Worcester, Winchester, and Norwich. Table 2 registers the weights examined and reported in the proceedings. All of them were recorded in pounds and were constructed during the reign of Henry VII unless marked to the contrary.

To demonstrate the magnitude of the discrepancies, let us examine the differences in grains between the Elizabethan Exchequer half-hundredweight and those found by the commissioners at the other five locations. The Norwich standard was 2679 grains greater than the Exchequer standard, which contained 392,000 grains. The standards of London, Exeter, and Worcester were each 1312 grains heavier than that of the Exchequer. The Winchester standard was 1313 grains less than the Exchequer standard. In this instance, therefore, the Exchequer standard represented the mean between the Winchester half-hundredweight and the three equal half-hundredweights of London, Exeter, and Worcester. As such, it might have been selected as the most representative weight in this group. But since there is no clue in the jury's proceedings as to which half-hundredweight was chosen, and since there is no mention made of the discrepancies in grains among weights of other denominations, it is impossible to know for sure which standards, either individually or within

10. In the "Seventh Annual Report of the Warden of the Standards," pp. 18-19, there is a comparison made between these weights and the imperial standards used in 1873.

Figure 5. Exchequer avoirdupois bell-shaped gunmetal weights, second series (1574) of Elizabeth I. The denominations are 7 and 28 pounds. (Photo. Science Museum, London)

a group, served as models for the almost five dozen sets of weights, both troy and avoirdupois, that were distributed in 1574 to the Exchequer and to the various local authorities. One can only assume that the Exchequer standards were selected.

Not many of the weights in this second series are extant; two are shown in Figure 5. Hastily constructed, poorly gauged, and devoid of artistic lettering and embossed reliefs, they were judged defective and recalled in 1582. A second jury of eighteen merchants and eleven goldsmiths was then impaneled. The new commissioners immediately condemned the weights made and verified by their predecessors, destroyed the faulty standards, and embarked upon the task of constructing and duly verifying new sets of avoirdupois and troy weights earmarked for the Exchequer and the principal cities and boroughs of England and Wales. The Exchequer primary reference standards were finished by the end of 1582, but it took six years to complete work on the copies. Of the copies, 43 sets were built for the same cities and boroughs to which Henry VII had sent his standards. The remainder were destined for the Tower of London, the Goldsmiths' Company, the Queen's Hospital, the borough of Brecon, and the towns of Marlborough, Boston, Colchester, Cardigan, Cardiff, Denbigh, Haverfordwest, Carmarthen, Carnarvon, Bala, and Flint. Local authorities sent representatives to London to claim their copies from the

Figure 6. Exchequer troy cup-shaped bronze weights, third series (1588) of Elizabeth I. The denominations are ⅛, ¼, ½, 1, 2, 4, 8, 16, 32, 64, 128, and 256 ounces. (Crown Copyright. Science Museum, London)

Figure 7. Exchequer avoirdupois bell-shaped and flat bronze weights, third series (1588) of Elizabeth I. The denominations of the bell-shaped weights are 1, 2, 4, 7, 14, 28, and 56 pounds. For the flat weights they are 2, 4, and 8 drams; 1, 2, 4, and 8 ounces; and 1, 2, 4, and 8 pounds. (Crown Copyright. Science Museum, London)

usher of the Exchequer. They paid this official a reasonable price for them and were instructed to destroy any local weights that did not agree with the new standards. Persons using illegal weights were liable to the penalties enunciated in the statutes of Henry VII. A proclamation that called for nationwide acceptance of the new standards was then posted in every market town. Additional copies of this directive were affixed on every church and read to parishioners twice a year for four consecutive years.

The Science Museum in London has a remarkable collection of these standards. Considering that they were used continually for 236 years, or from 1588 to 1824, each of the weights is in superior condition. All of them are bronze. They are magnificently engraved with a crown over the letter *E* and the date of 1582 (for the copies, a crown over *EL* and the date 1588), together with the year of the queen's reign and the denomination of each. The troy weights are a series of nested cups—the first such standards in English history (see Figure 6). When tested in 1873 by the warden of the standards, none of the weights up to 8 ounces (3840 grains) showed a deficiency of more than 1.09 grains. From 16 ounces to 128 ounces the worst deficiency was 5.04 grains on the 64-ounce weight (30,720 grains). For the largest cup weight—the one that would naturally take most of the wear, being the container for all the rest—the deficiency was 53.58 grains on 256 ounces (112,880 grains). It is important to note that the troy weights are inscribed, in Gothic letters, "oz tr". None are inscribed in troy pounds, so as to avoid confusion between them and the avoirdupois weights, which have "lb" inscribed in ornamental script.

The avoirdupois standards are in two sets: bell-shaped weights with solid loop handles on top, another first in English metrology; and flat, circular disk weights with slightly raised rims to allow all the weights to rest securely on top of each other in a conical pile, a form still favored for the smaller commercial weights (see Figure 7). In the reweighing of 1873, the 1- and 2-pound flat weights were deficient by 3.6 grains, and the 8-pound weight of the same set (56,000 grains) by a mere 12.05 grains. Of the bell-shaped weights, there was a 1-grain deficiency in the 1-pound standard and 116 grains in the half-hundredweight (392,000 grains).

Upon the completion of the new standards of troy and avoirdupois weight, two primary standards of length were constructed—the yard of 3 feet, or 36 inches, and the cloth ell of 45 inches (see Figure 1, page 76). Both measures were based on the bronze yard standard of Henry VII. Elizabeth's yard standard is a bronze rod, ½-inch square in section, the overall length being considered the yard. This "end standard" is only 0.01 inch shorter than the present imperial yard. The bar is subdivided by etched lines to show ½, ¼, ⅛, and ¹⁄₁₆ of a yard, the last (2¼ inches) representing the unit known as the "nail." The ell standard is a bronze rod, square in section similar to the yard, but 1¼ yards long overall. It is divided by coarse lines into ½, ¼, ⅛, and ¹/₁₆ parts, the latter being 2¹³/₁₆ inches. The entire rod is 45.04 inches by the current standard.

In addition there is a combined yard and ell "bed" measure or gauge, also shown in Figure 1. This is a broad, thick, heavy bar of bronze, rectangular in section, with an overall length of approximately 49 inches, a breadth of 1½ inches, and a thickness of 1 inch. One edge has been slotted out so as to contain the yard described above. The recessed length of this slot is marked out by coarse lines to indicate 3 feet. The first foot is subdivided into 12 inches and the first 2 inches into half-inches. The reverse edge on the other face is slotted out to take the ell rod. This bed, or matrix, measure was used as the principal gauge for testing local yards and ells until 1824, when a new yard standard was authorized by Parliament and the ell ceased to be a legal measure. All three standards are stamped at each end with a crown over the letter E.

The capacity measure standards of Elizabeth did not appear until 1601-2. Consisting of the Winchester bushel and gallon for dry commodities and of the ale gallon, quart, and pint, they served as the Exchequer primary reference standards until 1824. The Elizabethan Winchester measures were based on those of Henry VII and, like Henry's, they were determined by their troy weight content of wheat. All of these standards are richly decorated in the traditional Tudor fashion. In the special gauging of 1931, under the direction of J. E. Sears, then Deputy Warden of the Standards, the bushel was found to be 2148.28 cubic inches and the gallon, 268.97 cubic inches. These two measures, thus, were 3.47 and 0.54 cubic inches larger than the corresponding grain standards of Henry VII.

4

Epilogue

In the early years of the seventeenth century British metrology entered a new stage of historical evolution. Before the Stuart accession in 1603, metrological change had been achieved almost entirely through royal and parliamentary initiative. Royal decrees coupled with state and local enactments touched upon all three major aspects of weights and measures control—the units of measurement, the physical standards, and the officials entrusted with inspection, verification, and enforcement duties. In the long expanse of time from the early medieval to the early modern era, several hundred units of measurement had been defined or redefined. Continuous scientific, industrial, and technological progress had led to the production of physical standards of ever-increasing reliability and sophistication. A small corps of administrators had multiplied rapidly during these years and had acquired powers and duties that, especially in the political and economic realms, either rivaled or surpassed those bestowed on other representatives of the government bureaucracy. Despite all of the defects, errors, abuses, and other negative aspects connected with these developments, beneficial and highly significant change had taken place, especially in the manufacture of physical standards.

Nevertheless, to certain scientists, mathematicians, metrologists, and other concerned citizens after the Tudor era, the speed with which this change had occurred was far too slow. Decades, even centuries, were necessary to bring about reforms, and then all too often the results were impaired by ineptly conceived and poorly worded legislation, defects of various magnitudes in the standards, and inexperience, excessive competition, and corruption among those chosen to enforce the reforms. To many people what was needed was a restructuring of certain aspects of this metrology so that its inherent weaknesses would not reactivate the same problems in the future. But to others, such preventive proposals would never be successful, since the fundamental structure of the entire system was considered at fault. Instead of partially repairing a metrology that could never be made totally operable, these reformers pleaded that it would be far wiser to overhaul thoroughly the old system. Some of their plans utilized a decimal or other non-duodecimal scale for building unit proportions; some depended upon seconds pendulums or terrestrial measurements for establishing weights and measures standards; some eschewed the contemporary administrative framework and emphasized the need for an infinitely smaller and more professionally qualified inspectorate. But, regardless of their differences, each of them was motivated by a common belief—change could be speeded up. To them, it had to be speeded up.

Described best perhaps by the terms "contraction and redirection," the period from 1603 to 1824 witnessed a number of forces at work whose independent and sometimes correlative functions were responsible for the creation of the imperial system in the third decade of the nineteenth century. First, parliaments throughout this period labored to simplify and systematize the law of weights and measures. Antiquated statutes either were annulled or updated; statutory wording was modernized; and the scope of new and modified laws was made applicable throughout the British Isles. In those laws, unit standardization was given considerable attention and the total number of acceptable national units declined through various repeals and amalgamations. Foreign borrowings ceased and a massive assault was engineered on denominational, aggregate, product, and local variations with the ultimate aim of forcing acceptance of the London system. Of all these endeavors at unit standardization, that directed at local variations was the least successful.

Greater success was achieved in the manufacture of physical standards. By the middle of the eighteenth century, monumental advances were being made in many scientific and engineering fields. As theory and technology expanded, the need for more precise and sophisticated weighing and measuring apparatus, especially as regards the primary reference standards, grew more acute. Through parliamentary initiative the arsenal of standards bequeathed by the Tudors remained intact and many new and worthy additions, such as the Exchequer wine gallon (1707) of Queen Anne and the Winchester coal bushel (1730) of George II, enabled England to outdistance all other European nations in the number, precision, and accuracy of standards. Parliament was also instrumental in weeding out most part-time officials from the enormous bureaucratic network of weights and measures administration and in replacing them in the late eighteenth and early nineteenth centuries with a competent, experienced, full-time, professional inspectorate.

Assisting Parliament in all these tasks were government-sponsored commissions and independent scientists and scientific groups. Among the earliest was the Carysfort Committee, which from 1758 to 1760 supervised the construction of new standards for the troy pound and the yard. Legalized in 1824 as the primary reference standards for weight and length, they were destroyed by the Great Fire that ravaged the Houses of Parliament in 1834. Many of the recommendations of this committee concerning the proper specifications for unit standards, the reduction of the corpus of weights and measures law into one basic enactment, and the proper methods for checking and authorizing the legal standards became law after 1824.

The years between 1790 and 1824 were the most active in the areas of scientific experimentation, government-sponsored programs, and committee assignments, and numerous scientific societies began to provide themselves with weights and measures standards. A House of Commons committee formed in 1814 incorporated the results of this quarter-century of scientific experimentation into the English government's program of finding a more reliable basis for weights and measures standards. Researchers were achieving pioneering breakthroughs. As soon as the measurements of the earth began to be computed with some proficiency, the French proposed a standard measure of length based on the ten-millionth part of a quarter of the meridian. The Parisian savants expected such accurate precision that future measurements would not produce even a microscopic alteration in the length of their new meter. In England the pendulum was being given careful consideration. Since the second of time was determined by the motion of the earth, it was conceived that the length of the seconds pendulum in a given latitude would be an invariable quantity which could always be recovered or duplicated. Following extensive analysis, recommendations were offered for revising completely the structure of English linear and capacity measurement. Another parliamentary committee appointed in 1816 continued these investigations into the exact length of the seconds pendulum and compared the existing French and English standards of length. Finally, several committees meeting during the years 1819-23 drew up a series of proposals that would form the basis for the Imperial Weights and Measures Act of 1824. Additional committees were appointed thereafter (such as the Airy Commission of 1838-41, which considered the relationship between physical standards and measurement units on both the decimal and duodecimal scales; the Standards Commission of 1843, responsible for the construction of new linear and weight standards in 1855; and the Select Committee of 1862, which recommended the adoption of the metric system) and they were chiefly responsible for most of the changes that occurred in imperial metrology throughout the nineteenth and twentieth centuries.

Epilogue

The Imperial Weights and Measures Act revolutionized British metrology. All previous decrees, assizes, orders-in-council, ordinances, and statutes were repealed. Parliament broke irretrievably with the past and created a new law, one which in conjunction with the act of 1878 determined the fundamental course of British weights and measures until the seventh decade of the twentieth century. In the interim the number of measurement units was reduced sharply and those that were retained acquired further applications. Besides those traditional tasks performed by weights and measures officials, the inspectorate secured a wide range of additional duties such as regulating the measuring, sale, and transport of sand, ballast, coal, and other materials. In the first phase of the imperial system, or prior to the act of 1878, Parliament was interested principally in providing and maintaining the primary standards and in verifying copies of these for the use of local authorities. The major functions performed by inspectors (called examiners before 1835) were securing agreement between these standards and examining weights and weighing apparatus in factories and mercantile establishments. After 1878 Parliament moved in the direction of total regulation of weights and measures used in trade, and the power of inspectors increased proportionately. By the end of the nineteenth century these officials had gained a virtual monopoly over all aspects of weights and measures control. Drug and food controls took on a new importance in the law, with special emphasis on consumer protection against frauds and irregularities in retail and wholesale merchandising, in weighing and measuring procedures, and in methods of packaging. The act of 1824, together with the act of 1878 that reaffirmed the standards established in 1855, reduced the number of unit denominations still further, abolished the troy pound, and rejected all distinction between dry and liquid measures, were the proudest achievements in British metrological history until 1963.

During the first half of the twentieth century, however, so many radical changes occurred in various sectors of the economy that the nature and administrative pattern of weights and measures was altered considerably. Changes in the transportation and distribution of consumer goods and in marketing techniques—seen particularly in the expansion of multiple stores and supermarkets; technological and computational improvements that caused a mechanical revolution, especially in the design and use of such weighing and measuring devices as counting machines, fabric-measuring apparatus, churn filling instruments, and volumetric appliances; the rise of larger commercial, industrial, and mercantile conglomerates and a corresponding increase in the volume and variety of goods produced for retail and wholesale trade; and the prepackaged and frozen food phenomena—required a new weights and measures orientation and additional training for the inspection corps. There were slightly more than 500 inspectors by the turn of the century, and the 800-plus engaged in weights and measures control by the fifties found that many extraneous duties had been handed over to them—duties that were never performed or perhaps never even contemplated by earlier personnel. These new duties concerned the quality of food and other goods and the safety of the general public in matters outside police administration. They included the sampling and inspection of food and drugs for compositional qualities, the tests to be applied, the errors to be permitted, and the apparatus applicable in the mercantile sales of these goods. Similar duties were delegated to them in the areas of soil fertilizers and foods for cattle and poultry. Also added was the registration of dealers in and the storage of poisonous substances, especially insecticides and fungicides; the administration and enforcement of legislation regulating the closing hours of shops and the hours and conditions of employment of shop assistants; and a myriad of functions pertaining to the storage of explosives and petroleum, the falsity of description in the sale of goods, and the marking of imported fruits, vegetables, and meats on exposure for sale. In other words, inspectors were no longer merely "policemen" with punitive powers. They had become advisers of merchants and counselors of an increasingly consumption-conscious public.

For these reasons, by mid-century the weights and measures law had become substantially inadequate and outmoded. The Weights and Measures Act of 1963 removed these deficiencies and updated the law. This act constituted the basis of imperial metrology between 1963 and 1975.

Presently the British Isles are experiencing a period of metrological change and transition reminiscent of those decades in the early nineteenth century immediately preceding the introduction of the imperial system. On May 24, 1965, the president of the Board of Trade made a statement in the House of Commons accepting a request from industry for support by the government of industry's decision to implement a planned changeover to the metric system. It was recognized that the process of change would be gradual, and a time span of ten years was proposed for the industrial transition to the new system. By 1976 the changeover was proceeding according to initial plans, although there was some resistance in certain private social and economic sectors. The pharmaceutical industry was the first to convert; some others, such as construction, engineering, cable manufacturing, paper, printing, and photography, were completing their conversions. In the nonindustrial sector, the Consumer Council, the Shippers' Council, the Food Manufacturers' Federation, and the Civil Aviation Authorities coordinated a changeover with industry.

Epilogue

On May 14, 1968, the Standing Joint Committee on Metrication, with the support of the Council of the Confederation of British Industry, the British Standards Institution, the Royal Society, and the Council of Engineering Institutions, implemented the 1965 statement and recommended that the government establish the end of 1975 as the guideline date for the adoption of the metric system by the country as a whole, it being understood that programs for the changeover were to be devised by those concerned, sector by sector within the economy, against this guideline date. These programs could, on detailed examination of the problems involved, aim at an earlier or later date. Second, planning bodies representing all interests concerned with a particular sector of the economy were established to identify the problems, and to prepare sector programs of change against the guideline date of 1975. A Metrication Board was established under the sponsorship of a central department of government with its own full-time staff to oversee, stimulate, and coordinate the sector planning bodies. The board was thus the central planning agency, and was responsible for preparing the public for the change. Finally, enabling legislation, specifying the metric units to be used and permitting the necessary changes to be made in all sectors by statutory instrument, was prepared.

Despite some lingering resistance to metrication, Britain has brought to a close almost two millennia of weights and measures developments that culminated in the imperial system. A new metrological law is now being written.

Reference Matter

Appendix A

Weights and Measures of Merchandise in the English Import and Export Trade, ca. 1500 to ca. 1800

These alphabetical lists give the principal imports and exports of England over a three-hundred-year period, together with the weights or measures by which the articles of merchandise were bulk rated at the government's customs stations. The products were compiled from the available port records, merchants' manuals, books of rates and prices, and pertinent manuscripts and monographs. Spellings of the articles of merchandise have been modernized when a modern equivalent exists. Words in parentheses following the names of articles are modifiers that appeared in the original sources; words in brackets are interpolations, which have been added when necessary to clarify the meaning of names.

All of the weights and measures used at the customs stations conformed to dimensions or specifications outlined in statute law or in specific government directives. In the second column the value of each unit and its metric equivalent are provided wherever necessary.

List of Abbreviations

dkl	=	dekaliter or dekaliters
g	=	gram or grams
hl	=	hektoliter or hektoliters
kg	=	kilogram or kilograms
l	=	liter or liters
m	=	meter or meters

Appendix A

IMPORTS

Description of Merchandise	Bulk Rated
acacia [a gum arabic]	avoirdupois pound (453.592 g)
acorns	avoirdupois pound (453.592 g)
acorus [a rushlike herb]	avoirdupois pound (453.592 g)
adzes [for coopers]	dozen
agaric [the dried fruit body of a mushroom]	avoirdupois pound (453.592 g)
agates (large) [fine-grained chalcedony]	each
aglets [ornamental pins or cords worn on clothing]	gross
agnus castus [a stimulant drug]	avoirdupois pound (453.592 g)
alabaster [fine-textured gypsum]	load (weight unspecified)
alkali [a soluble salt]	hundredweight of 112 avoirdupois pounds (50.802 kg)
alkanet root [a red dyestuff]	avoirdupois pound (453.592 g)
almain rivets [a flexible light armor]	each
almonds	hundredweight of 108 avoirdupois pounds (48.988 kg); bale of 3 hundredweight (146.964 kg)
aloes [the plant]	hundredweight of 100 avoirdupois pounds (45.359 kg)
alphabets	set
alpist [seeds]	hundredweight of 112 avoirdupois pounds (50.802 kg)
alum [an emetic or astringent]	hundredweight of 108 avoirdupois pounds (48.988 kg)
amber [a polish]	avoirdupois pound (453.592 g); mast of 2½ avoirdupois pounds (1.134 kg)
ambergris [a fixative in perfumery]	avoirdupois ounce (28.350 g)
ammoniac [an aromatic gum resin]	avoirdupois pound (453.592 g)
anchors and locks	shock of 60 in number
anchovies	barrel of 30 avoirdupois pounds (13.608 kg)
andirons	pair
andlets [the meaning is not certain]	avoirdupois pound (453.592 g); hundredweight of 112 avoirdupois pounds (50.802 kg)
angelica [an herb]	avoirdupois pound (453.592 g)
aniseed [an herb]	hundredweight of 112 avoirdupois pounds (50.802 kg)

English Import and Export Items

Imports (continued)

Description of Merchandise	Bulk Rated
annatto [a reddish dyestuff]	avoirdupois pound (453.592 g)
antimony [a mineral used in alloys]	hundredweight of 100 avoirdupois pounds (45.359 kg)
anvils	hundredweight of 112 avoirdupois pounds (50.802 kg)
apples	bushel (35.238 l); barrel of 3 bushels (ca. 1.06 hl)
aqua vitae [a strong liquor; brandy, whiskey]	barrel (ca. 1.36 hl); hogshead (capacity or weight unspecified)
argal [a dry dung fuel]	hundredweight of 112 avoirdupois pounds (50.802 kg)
aristolochia [an herb]	avoirdupois pound (453.592 g)
armor (old)	hundredweight of 112 avoirdupois pounds (50.802 kg)
arras [a high-warp tapestry]	ell of 45 inches (1.143 m)
arrows	gross
arsenic	avoirdupois pound (453.592 g)
asafetida [a gum resin]	avoirdupois pound (453.592 g); hundredweight of 100 avoirdupois pounds (45.359 kg)
asarum root [an herb]	avoirdupois pound (453.592 g)
ashes	barrel (ca. 1.48 hl); last of 12 barrels (ca. 17.76 hl)
asphalt	avoirdupois pound (453.592 g)
augers	dozen; gross
awl blades or hafts	thousand of 1,000 in number
axes	dozen
babies [puppets for children]	gross
bacon	flitch (a side of cured hog meat); hundredweight of 112 avoirdupois pounds (50.802 kg)
badger skins	each
bags	dozen
baize [a coarsely woven woolen or cotton fabric]	yard
balances	gross
balaustia [a drug]	avoirdupois pound (453.592 g)
balks [roughly squared pieces of timber]	hundred of 120 in number
balls	dozen; gross; thousand of 1,000 in number
balsam [an aromatic substance]	avoirdupois pound (453.592 g)
bands	each; dozen

Appendix A

Imports (continued)

Description of Merchandise	Bulk Rated
bandstrings [for ruffs or collars]	dozen knots (quantity unspecified)
bankers (of verdure) [bench or chair coverings]	dozen pieces of 24 yards (ca. 21.95 m) in length and 7 quarters (ca. 1.60 m) in breadth each
barbers' aprons	piece (length and breadth unspecified)
barilla [an impure sodium carbonate]	barrel of 2 hundredweight (101.604 kg)
barlings or firepoles	hundred of 120 in number
barwood [a hard red dyewood]	ton of 2,240 avoirdupois pounds (1,016.040 kg)
Basel leather	dozen
basins (of latten)	avoirdupois pound (453.592 g)
basket-rods	bundle (quantity unspecified)
baskets	each
bast or straw hats	dozen; thousand of 1,000 in number
bast ropes	each; shock of 60 in number; bundle of 10 in number; hundredweight of 112 avoirdupois pounds (50.802 kg)
battery [kettles]	hundredweight of 112 avoirdupois pounds (50.802 kg)
bayberries	hundredweight of 112 avoirdupois pounds (50.802 kg)
beads	avoirdupois pound (453.592 g); hundred of 100 in number; gross; thousand of 1,000 in number
beans	bushel (35.238 l)
bear skins	each
beaver skins	timber of 40 in number
beds	each
beef	barrel of 32 wine gallons (ca. 1.21 hl); puncheon (capacity or weight unspecified)
beer	last of 12 barrels (ca. 19.92 hl)
bell metal	hundredweight of 112 avoirdupois pounds (50.802 kg)
bellows	pair
bells	avoirdupois pound (453.592 g); hundredweight of 112 avoirdupois pounds (50.802 kg); dozen pairs; gross
belts	each; dozen; gross
benjamin [the plant]	avoirdupois pound (453.592 g); hundredweight of 100 avoirdupois pounds (45.359 kg)

English Import and Export Items

Imports (continued)

Description of Merchandise	Bulk Rated
bezoar stones [a medicine or pigment]	avoirdupois ounce (28.350 g)
birdlime [an adhesive substance]	hundredweight of 112 avoirdupois pounds (50.802 kg)
bitumen [an asphalt used as a cement or mortar]	avoirdupois pound (453.592 g)
blacking or lampblack [a fine bulky black soot]	hundredweight of 112 avoirdupois pounds (50.802 kg)
black latten rolls	hundredweight of 112 avoirdupois pounds (50.802 kg)
blankets	piece (length and breadth unspecified)
boards	each; hundred of 120 in number; thousand of 1,200 in number
boats (4 or 6 oars)	each
bodkins [sharp pointed tools]	gross; thousand of 1,000 in number
bolting cloth [a fine silk fabric]	piece of 24 yards (ca. 21.95 m) in length and 7 quarters (ca. 1.60 m) in breadth
bombazine [a silk fabric in twill weave]	piece of 24 yards (ca. 21.95 m) in length and 7 quarters (ca. 1.60 m) in breadth
bonnets	each; dozen
books (unbound)	basket (quantity unspecified); maund of 40 reams
boom spars	hundred of 120 in number
boots	dozen pairs
borax [a sodium borate]	avoirdupois pound (453.592 g)
bosses [for bridles]	gross
bottanos [blue linen]	piece of 24 yards (ca. 21.95 m) in length and 7 quarters (ca. 1.60 m) in breadth
bottles	dozen; gross
boultel [a cloth]	piece of 24 yards (ca. 21.95 m) in length and 7 quarters (ca. 1.60 m) in breadth; dozen; bale of 20 pieces
bows	each; dozen
bowstaves	bundle (quantity unspecified); hundred of 120 in number; last of 720 in number
bowstrings	dozen
boxes	dozen; shock of 60 in number; gross
boxwood	ton of 2,240 avoirdupois pounds (1,016.040 kg)
Brabant cloth	half or whole piece, the latter of 24 yards (ca. 21.95 m) in length and 7 quarters (ca. 1.60 m) in breadth
bracelets	gross

Imports (continued)

Description of Merchandise	Bulk Rated
brandy	gallon (3.785 l); tun (ca. 9.54 hl)
brass	hundredweight of 100 avoirdupois pounds (45.359 kg)
brass lamps	dozen
brass lavercocks	avoirdupois pound (453.592 g)
brass pile weights	avoirdupois pound (453.592 g)
brass trumpets	dozen
brazilwood	hundredweight of 112 avoirdupois pounds (50.802 kg)
brick stones and tiles	thousand of 1,200 in number
bridle bits	dozen
bridle reins	each
bridles	dozen
brimstone [sulfur]	hundredweight of 112 avoirdupois pounds (50.802 kg)
bristles	dozen avoirdupois pounds (5.443 kg)
broaches	gross
brushes	dozen; gross
Brussels cloth	piece of 24 yards (ca. 21.95 m) in length and 7 quarters (ca. 1.60 m) in breadth
buckles	gross
buckram [a fine linen or cotton fabric]	piece of 15 yards (ca. 13.72 m) in length and 7 quarters (ca. 1.60 m) in breadth; roll (dimensions unspecified)
budge [a fur prepared from lambskin and dressed with wool]	each; dozen; hundred of 100 in number
buff hides [wild ox hides]	each
buffins [coarse cloth]	each
bugle [the plant]	avoirdupois pound (453.592 g)
bullions [knobs or bosses] (for purses)	gross
bulrushes [the plant]	load of 63 bundles
burrs [rock for millstones]	hundred of 120 in number
buskins (of leather) [a foot covering]	dozen pairs
bustians [a cotton fabric]	piece of 24 yards (ca. 21.95 m) in length and 7 quarters (ca. 1.60 m) in breadth
butter	barrel of 32 ale gallons (ca. 1.48 hl)
buttons	dozen; gross; great gross of 1,728 in number
cabbages	hundred of 100 in number
cabinets	each
cables [strong, heavy ropes]	stone of 14 avoirdupois pounds (6.350 kg); hundredweight of 112 avoirdupois pounds (50.802 kg)

Imports (continued)

Description of Merchandise	Bulk Rated
Caen stone [a yellowish limestone]	ton of 2,240 avoirdupois pounds (1,016.040 kg)
calaber [a squirrel fur]	each; timber of 40 in number
calamus [a reed pen]	hundred of 100 in number
calf skins	each
calico [a white cotton fabric]	piece of 24 yards (ca. 21.95 m) in length and 7 quarters (ca. 1.60 m) in breadth
camel's hair	avoirdupois pound (453.592 g)
camlet [a fabric of camel's hair or Angora wool]	yard
camomile flowers [used as an antispasmodic or diaphoretic]	avoirdupois pound (453.592 g)
camphor	avoirdupois pound (453.592 g)
camwood [a hard red wood]	ton of 2,240 avoirdupois pounds (1,016.040 kg)
candle plates	dozen
candles	avoirdupois pound (453.592 g)
candlesnuffers	dozen
candlesticks	avoirdupois pound (453.592 g); dozen
candlewicks	hundredweight of 112 avoirdupois pounds (50.802 kg)
canella alba [a tonic]	avoirdupois pound (453.592 g)
canes	dozen; thousand of 1,000 in number
cans [small containers, generally of wood]	shock of 60 in number
cant spars [poles]	hundred of 120 in number
canvas	piece of 24 yards (ca. 21.95 m) in length and 7 quarters (ca. 1.60 m) in breadth; fardel (dimensions unspecified); bale (dimensions unspecified); bolt (dimensions unspecified); hundred ells of 45 inches (1.143 m) each; hundredweight of 100 avoirdupois pounds (45.359 kg)
capers [a condiment]	avoirdupois pound (453.592 g)
cap hooks	gross
capravens [the meaning is not certain]	hundred of 120 in number
caps	dozen
caraway seeds	hundredweight of 112 avoirdupois pounds (50.802 kg)
cardamoms [an aromatic fruit]	hundredweight of 112 avoirdupois pounds (50.802 kg)
cards	dozen; dozen pairs; twelve dozen pairs
carpets	each; square yard; dozen

Appendix A

Imports (continued)

Description of Merchandise	Bulk Rated
carts	gross
carving knives and tools	dozen; stock of 60 in number; case (quantity unspecified)
cases	dozen; gross
caskets	dozen
cassia fistula [dried pods of the drumstick tree]	hundredweight of 112 avoirdupois pounds (50.802 kg)
cassia lignea [cinnamon in thick flat pieces]	avoirdupois pound (453.592 g)
castor [the bean]	avoirdupois pound (453.592 g)
cat skins	hundred of 100 in number
cauls [women's netted close-fitting caps]	dozen
caviar	hundredweight of 112 avoirdupois pounds (50.802 kg)
cedarwood	ton of 2,240 avoirdupois pounds (1,016.040 kg)
chafing dishes	dozen
chains	dozen
chairs	each
chalk	hundredweight of 112 avoirdupois pounds (50.802 kg)
chandeliers	avoirdupois pound (453.592 g); dozen
cheese	hundredweight of 112 avoirdupois pounds (50.802 kg)
cherries	hundredweight of 112 avoirdupois pounds (50.802 kg)
chessboards	dozen
chessmen	gross
chests	each; nest of 3 in number
chimney backs	each
China peas [the meaning is not certain]	avoirdupois pound (453.592 g)
chisels	half or whole dozen
chocolate	avoirdupois pound (453.592 g)
cider	tun of 252 gallons (ca. 9.54 hl)
cinnamon	hundredweight of 108 avoirdupois pounds (48.988 kg)
cisterns (of tin)	avoirdupois pound (453.592 g)
citrons [citrus fruit or melons]	dozen
civet [a perfume]	avoirdupois ounce (28.350 g)
clapboards [for staves]	hundred of 120 in number
cloaks	each

English Import and Export Items

Imports (continued)

Description of Merchandise	Bulk Rated
cloth	piece of 24 yards (ca. 21.95 m) in length and 7 quarters (ca. 1.60 m) in breadth; yard; ell of 45 inches (1.143 m)
cloves	avoirdupois pound (453.592 g); hundredweight of 100 avoirdupois pounds (45.359 g)
cobalt [for alloys]	avoirdupois pound (453.592 g)
cochineal [a red dyestuff]	avoirdupois pound (453.592 g)
codfish	hundred of 124 in number; last of 496 in number; barrel of 32 ale gallons (ca. 1.48 hl)
coffee	hundredweight of 112 avoirdupois pounds (50.802 kg)
coffers	each; nest of 3 in number; dozen
coifs [small hooded caps]	dozen
collars (for dogs)	dozen
coloquintida [colocynth; a cathartic]	avoirdupois pound (453.592 g)
combs	pair; dozen; gross; box (quantity unspecified); case (quantity unspecified); avoirdupois pound (453.592 g)
comfits [a coated confection]	avoirdupois pound (453.592 g)
compasses	each; dozen
copper	hundredweight of 112 avoirdupois pounds (50.802 kg)
copperas [a green dye]	avoirdupois pound (453.592 g); hundredweight of 112 avoirdupois pounds (50.802 kg); pipe (capacity or weight unspecified)
copper quills [the meaning is not certain]	avoirdupois pound (453.592 g)
copper rings	dozen
coral, red or white [a dye]	avoirdupois pound (453.592 g)
corbels [architectural supports]	each
cordage [ship cords or ropes]	stone of 14 avoirdupois pounds (6.350 kg); hundredweight of 112 avoirdupois pounds (50.802 kg)
cordovans [leather goods]	dozen
coriander seeds [an herb]	hundredweight of 112 avoirdupois pounds (50.802 kg)
cork	hundredweight of 112 avoirdupois pounds (50.802 kg); barrel (capacity or weight unspecified); last (capacity or weight unspecified)

Appendix A

Imports (continued)

Description of Merchandise	Bulk Rated
corn	bushel (35.238 l); quarter (ca. 2.82 hl)
costus oil [a perfume]	avoirdupois pound (453.592 g)
cotton	yard; ell of 45 inches (1.143 m); hundredweight of 100 avoirdupois pounds (45.359 kg)
counters [instruments used in reckoning]	each; nest of 3 in number
coverlets	each
cowhage [a vermifuge]	avoirdupois pound (453.592 g)
cows	each
cranberries	gallon (capacity or weight unspecified)
crêpe [a lightweight fabric]	ell of 45 inches (1.143 m); piece of 24 yards (ca. 21.95 m) in length and 7 quarters (ca. 1.60 m) in breadth
crossbow lathes and thread	avoirdupois pound (453.592 g)
crossbow racks	each
crosses (of stone)	hundred of 100 in number
cubebs [a medicine]	avoirdupois pound (453.592 g)
cullen gold or silver [from Cologne]	mast of 2½ avoirdupois pounds (1.134 kg)
cullen hemp [from Cologne]	stone of 16 avoirdupois pounds (7.257 kg); hundredweight of 112 avoirdupois pounds (50.802 kg)
cullen thread [from Cologne]	bale (quantity or weight unspecified)
culm [refuse coal]	ton of 2,240 avoirdupois pounds (1,016.040 kg)
cumin [aromatic flavoring seeds]	hundredweight of 112 avoirdupois pounds (50.802 kg); bale (capacity or weight unspecified)
currants [seedless raisins]	hundredweight of 112 avoirdupois pounds (50.802 kg)
curtain rings	avoirdupois pound (453.592 g)
cushions	dozen
cuttlebones [for making polishing powder]	thousand of 1,000 in number
daggers	dozen; gross
damask [a firm lustrous satin fabric]	yard
Danish leather	dozen skins
dates	hundredweight of 112 avoirdupois pounds (50.802 kg)
deals [boards of fir or pine]	each; hundred of 120 in number

English Import and Export Items

Imports (continued)

Description of Merchandise	Bulk Rated
desks	each; dozen
dials [mariners' compasses]	dozen
diaper [a white linen or cotton fabric] napkins	dozen
diaper tablecloths	yard
diaper toweling	piece of 24 yards (ca. 21.95 m) in length and 7 quarters (ca. 1.60 m) in breadth
dimity [a sheer cotton fabric]	yard
dittany [an aromatic herb]	avoirdupois pound (453.592 g)
dog chains	gross
dogs (of earthenware)	gross
dornick [a coarse damask of wool and silk]	piece of 28 yards (ca. 25.60 m) in length and 7 quarters (ca. 1.60 m) in breadth; yard; ell of 45 inches (1.143 m)
doubles [harness plates]	each; bundle (quantity unspecified); shock of 60 in number
dowlas [a coarse linen cloth]	piece of 24 yards (ca. 21.95 m) in length and 7 quarters (ca. 1.60 m) in breadth
down [soft fluffy feathers]	avoirdupois pound (453.592 g)
dressers	each
drinking glasses	dozen
druggets [a cotton-warped and wool-filled rug]	each
earings [sail lines]	gross
earthenware tiles	thousand of 1,200 in number
ebony wood	hundredweight of 112 avoirdupois pounds (50.802 kg)
eels	barrel of 30 gallons (ca. 1.14 hl); last of 12 barrels (ca. 13.68 hl)
eggs	hundred of 120 in number
egrets [any of the various herons]	dozen
elephants' teeth	hundredweight of 112 avoirdupois pounds (50.802 kg)
emery stones	hundredweight of 112 avoirdupois pounds (50.802 kg)
empty barrels (for grocers)	dozen
enamel	avoirdupois pound (453.592 g)
euphorbium [a yellow or brownish gum resin]	avoirdupois pound (453.592 g)
falcons	each
fans	avoirdupois pound (453.592 g); each; dozen
fawn skins	each

Imports (continued)

Description of Merchandise	Bulk Rated
featherbeds	each
feathers	avoirdupois pound (453.592 g); hundredweight of 112 avoirdupois pounds (50.802 kg)
felts (for cloaks)	each
fiddles (toy)	dozen
figs	piece of 4 quarterns (50.802 kg); sort of 3 pieces; tapnet of 20 to 30 avoirdupois pounds (9.072 to 13.608 kg); hundredweight of 112 avoirdupois pounds (50.802 kg)
figuretto [figured cloth]	yard
files	gross
filings (of iron)	hundredweight of 112 avoirdupois pounds (50.802 kg)
fire shovels	dozen
fitches [the fur or pelt of the fitch]	each; timber of 40 in number
flagons	dozen
Flanders bricks	thousand of 1,000 in number
flannel	yard; ell of 45 inches (1.143 m)
flasks	dozen
flax [the bast fiber]	dozen avoirdupois pounds (5.443 kg); hundredweight of 112 avoirdupois pounds (50.802 kg); bale (weight unspecified); pack of 240 avoirdupois pounds (108.862 kg)
fleams [sharp lancets used for bloodletting]	each
Flemish cloth	piece of 24 yards (ca. 21.95 m) in length and 7 quarters (ca. 1.60 m) in breadth
flocks [woolen or cotton refuse]	hundredweight of 112 avoirdupois pounds (50.802 kg)
Florence woolen cloth	yard
flutes	gross
foin skins [from the stone marten]	dozen
foin tails [from the stone marten]	each
fox skins	each
frankincense [a gum resin]	hundredweight of 112 avoirdupois pounds (50.802 kg)
French carpets	yard; ell of 45 inches (1.143 m)
French woolen cloth	piece of 24 yards (ca. 21.95 m) in length and 7 quarters (ca. 1.60 m) in breadth
frizado [a coarse woolen cloth]	piece of 24 yards (ca. 21.95 m) in length and 7 quarters [ca. 1.60 m) in breadth

English Import and Export Items 115

Imports (continued)

Description of Merchandise	Bulk Rated
frying pans	hundredweight of 112 avoirdupois pounds (50.802 kg)
furs	each; dozen; timber of 40 in number; hundred of 100 in number
fustians [a strong cotton and linen fabric]	piece of 13 ells (14.859 m) in length and 7 quarters (ca. 1.60 m) in breadth; bale of 40 or 45 half-pieces
fustic [a yellow dye]	hundredweight of 112 avoirdupois pounds (50.802 kg)
gadza [a cloth]	yard; ell of 45 inches (1.143 m)
galangal [a pungent aromatic spice]	avoirdupois pound (453.592 g)
galbanum [an aromatic gum resin]	hundredweight of 112 avoirdupois pounds (50.802 kg)
galley pots	dozen; hundred of 100 in number
galls [the meaning is not certain]	hundredweight of 112 avoirdupois pounds (50.802 kg)
gamboge [a gum resin]	avoirdupois pound (453.592 g)
garlic	hundred bunches of 2,500 in number
garnets [a semiprecious stone and an abrasive]	avoirdupois pound (453.592 g)
garrons [spike nails]	hundred of 120 in number
garters	pair; dozen pairs
gauntlets [reinforced gloves used with armor]	pair; dozen
gentian [a tonic and stomachic]	avoirdupois pound (453.592 g)
gimlets [small woodworking tools with a screw point]	dozen
ginger	avoirdupois pound (453.592 g); hundredweight of 100 avoirdupois pounds (45.359 kg)
ginseng [a Chinese herb]	avoirdupois pound (453.592 g)
girdles	dozen; gross
girds (of iron for puncheons or pipes)	hundredweight of 112 avoirdupois pounds (50.802 kg)
girth web [a band or strap]	gross
gitterns [stringed, guitarlike instruments]	each; dozen
glasses	dozen; hundred of 100 in number; gross; barrel (quantity unspecified); case or chest (quantity unspecified); web or wey of 60 cases of uncertain weight
globes	pair
gloves	dozen pairs; gross

Appendix A

Imports (continued)

Description of Merchandise	Bulk Rated
glue	hundredweight of 112 avoirdupois pounds (50.802 kg)
goat skins	dozen
gold foil or paper	gross
granilla [the meaning is not certain]	avoirdupois pound (453.592 g)
gray skins [from the badger]	each; timber of 40 in number
grindstones [millstones]	hundred of 120 in number
grograms [a coarse, loosely woven fabric of silk and wool]	yard
gum arabic [a water-soluble gum obtained from the acacia]	hundredweight of 112 avoirdupois pounds (50.802 kg)
gum dragon [tragacanth; a thickening agent]	avoirdupois pound (453.592 g)
gum elemi [a fragrant oleoresin]	avoirdupois pound (453.592 g)
gunpowder	hundredweight of 100 avoirdupois pounds (45.359 kg)
guns	each
gypsum	ton of 2,240 avoirdupois pounds (1,016.040 kg)
haddock	last of 12 barrels (ca. 17.76 hl)
hair	avoirdupois pound (453.592 g); hundredweight of 112 avoirdupois pounds (50.802 kg)
hair sieves	gross
halberds [combination long-handled battle axes and pikes]	each
hammers	dozen
hampers	nest of 3 in number
hams	hundredweight of 112 avoirdupois pounds (50.802 kg)
handkerchiefs	dozen
hanging locks	gross
harness corslets, cuirasses, or morions	each
harpstrings	gross; box (quantity unspecified)
hatbands	gross
hats	each; dozen
hawks	each
hawks' hoods	gross
heather [the plant]	hundredweight of 112 avoirdupois pounds (50.802 kg)

Imports (continued)

Description of Merchandise	Bulk Rated
hellebore [a medicinal herb]	avoirdupois pound (453.592 g)
hemp	hundredweight of 112 avoirdupois pounds (50.802 kg); sack of 3 hundredweight (152.406 kg); last (weight unspecified)
herrings	cade of 500 to 1,000 in number; last of 12,000 in number
hides	each
hilts (for daggers or swords)	dozen; gross
Holland cloth	piece of 24 yards (ca. 21.95 m) in length and 7 quarters (ca. 1.60 m) in breadth
hones [fine-grit stones for sharpening cutting instruments]	hundred of 100 in number
honey	barrel of 31½ gallons (ca. 1.19 hl)
hooks	dozen; gross
hoops (for barrels)	thousand of 1,000 in number
hops	hundredweight of 112 avoirdupois pounds (50.802 kg); pocket of 1½ to 2 hundredweight (76.203 to 101.604 kg); poke (weight unspecified); sack (weight unspecified)
horns	dozen; thousand of 1,000 in number
horse combs	dozen
horses	each
horseshoes	dozen
hose (of silk)	pair
hourglasses	dozen
huskins (for fletchers) [arrow-making implements]	each
imperlings [the meaning is not certain]	dozen
indigo [a blue dye]	avoirdupois pound (453.592 g)
ink	hundredweight of 112 avoirdupois pounds (50.802 kg)
inkhorns and pens	dozen; gross
inkle [a colored linen tape or braid]	avoirdupois pound (453.592 g); dozen avoirdupois pounds (5.443 kg); hundredweight of 112 avoirdupois pounds (50.802 kg)
instruments (for surgeons)	each; dozen
iron	stone of 14 avoirdupois pounds (6.350 kg); hundredweight of 112 avoirdupois pounds (50.802 kg); bundle (quantity or weight unspecified); ton of 2,240 avoirdupois pounds (1,016.040 kg)

Appendix A

Imports (continued)

Description of Merchandise	Bulk Rated
iron bands (for kettles)	hundredweight of 112 avoirdupois pounds (50.802 kg)
iron chests	each
iron pots	dozen
iron rings (for chains)	avoirdupois pound (453.592 g)
iron stoves	each
iron wire	hundredweight of 112 avoirdupois pounds (50.802 kg)
isinglass [a clarifying agent]	avoirdupois pound (453.592 g)
ivory	avoirdupois pound (453.592 g)
jalap [a powdered drug]	avoirdupois pound (453.592 g)
jasper stones [reddish quartz used for polishing]	hundred of 120 in number
javelin heads and staves	each; dozen
jennets [the skins or pelts]	each; timber of 40 in number
jet [a velvet-black mineral]	avoirdupois pound (453.592 g)
jew's trumps [jew's harps; small lyre-shaped instruments]	gross
juniper berries	hundredweight of 112 avoirdupois pounds (50.802 kg)
kells [women's net caps]	dozen
kettles	hundredweight of 112 avoirdupois pounds (50.802 kg)
kid skins	each
kists [chests or coffins]	each
knives	half and whole dozen; dicker of 10 in number; gross
lace	twelve yards; troy pound (373.242 g) or avoirdupois pound (453.592 g)
ladles	avoirdupois pound (453.592 g); hundredweight of 112 avoirdupois pounds (50.802 kg)
lambs	each
lampblack [a fine bulky black soot]	hundredweight of 112 avoirdupois pounds (50.802 kg)
lampreys [eels]	each
lanterns	dozen
lapis lazuli [a semiprecious blue stone]	troy pound (373.242 g)
lard	hundredweight of 112 avoirdupois pounds (50.802 kg)
lasts (for shoemakers)	dozen

English Import and Export Items 119

Imports (continued)

Description of Merchandise	Bulk Rated
latten [an alloy of or resembling brass]	hundredweight of 112 avoirdupois pounds (50.802 kg)
latten basins	hundredweight of 112 avoirdupois pounds (50.802 kg)
latten shoehorns	dozen
latten wire	hundredweight of 112 avoirdupois pounds (50.802 kg)
launders [water troughs]	pair
lawn [a sheer plain-woven cotton or linen fabric]	piece of 18 yards (ca. 16.46 m) in length and 7 quarters (ca. 1.60 m) in breadth
lead ore	hundredweight of 112 avoirdupois pounds (50.802 kg); ton of 2,240 avoirdupois pounds (1,016.040 kg)
leather hides or skins	each; dozen
leaves of gold	hundred of 100 in number
lemons (pickled)	pipe (weight unspecified)
lemon water	tun of 252 gallons (ca. 9.54 hl)
lentils	bushel (35.238 l)
leopard skins	each
licorice	bale of 2 hundredweight (101.604 kg)
lignaloes [agalloch; a perfume]	avoirdupois pound (453.592 kg)
lignum vitae [the wood]	hundredweight of 112 avoirdupois pounds (50.802 kg)
lime (for dyers)	barrel (weight unspecified)
linen cloth	piece of 24 yards (ca. 21.95 m) in length and 7 quarters (ca. 1.60 m) in breadth; yard; ell of 45 inches (1.143 m); hundred ells; bolt (dimensions unspecified); roll (dimensions unspecified)
lines (for fishing)	each
lings [deep-sea fish]	hundred of 124 in number
linseed	bushel (35.238 l)
linseed cakes	hundredweight of 112 avoirdupois pounds (50.802 kg)
lint [flax]	stone of 14 avoirdupois pounds (6.350 kg); last (weight unspecified)
litmus [a dye]	hundredweight of 112 avoirdupois pounds; barrel (weight unspecified)
lockets [small cases]	gross
lockram [a coarse plain-woven linen cloth]	piece of 24 yards (ca. 21.95 m) in length and 7 quarters (ca. 1.60 m) in breadth
locks	dozen; gross

Appendix A

Imports (continued)

Description of Merchandise	Bulk Rated
logwood	hundredweight of 112 avoirdupois pounds (50.802 kg)
long pepper [a condiment]	avoirdupois pound (453.592 g); hundredweight of 108 avoirdupois pounds (48.988 kg)
looking glasses [mirrors]	dozen
lormery [metalware]	hundredweight of 112 avoirdupois pounds (50.802 kg)
lupines [edible seeds]	hundredweight of 112 avoirdupois pounds (50.802 kg)
lures (for hawks)	each
lutes	each; dozen
lute strings	gross
mace [an aromatic spice]	avoirdupois pound (453.592 g); hundredweight of 112 avoirdupois pounds (50.802 kg)
madder [a dye]	hundredweight of 112 avoirdupois pounds (50.802 kg); bale (weight unspecified)
mahogany	ton of 2,240 avoirdupois pounds (1,016.040 kg)
manna [a laxative and demulcent]	avoirdupois pound (453.592 g)
maps (printed)	ream of 20 quires or 500 sheets
margarite [a mineral related to mica]	avoirdupois ounce (28.350 g)
marking stones	avoirdupois pound (453.592 g)
marmalade	avoirdupois pound (453.592 g)
martens [the pelts or skins]	timber of 40 in number
masks	dozen
mastic [an aromatic resin]	hundredweight of 112 avoirdupois pounds (50.802 kg)
masts	each
match (for guns)	avoirdupois pound (453.592 g)
mats	each
mead [a liquor of fermented honey]	gallon (3.785 l)
meal (wheat or rye)	last of 12 barrels (ca. 17.76 hl)
melting pots (for goldsmiths)	thousand of 1,000 in number
metheglin [mead]	hogshead (capacity unspecified)
millstones [for grinding grain]	each
minikins [thin gut treble strings for viols or lutes]	gross
miniver [a whitish fur from the vair, ermine, or rabbit]	each

English Import and Export Items

Imports (continued)

Description of Merchandise	Bulk Rated
minks [the fur]	timber of 40 in number
mithridate [an electuary used as a remedy against poison]	avoirdupois pound (453.592 g)
moden (starch)	hundredweight of 112 avoirdupois pounds (50.802 kg)
mohair	hundredweight of 112 avoirdupois pounds (50.802 kg)
molasses	tun of 252 gallons (ca. 9.54 hl)
mole skins	dozen
morris pikes [large pikes used by foot soldiers]	dozen
mortars and pestles	dozen
mules	each
musk [a fixative in perfumes]	avoirdupois ounce (28.350 g)
muslins [plain-woven cotton fabrics]	piece of 14 ells (16.002 m) in length and 7 quarters (ca. 1.60 m) in breadth
mustard seed	hundredweight of 112 avoirdupois pounds (50.802 kg); barrel (weight unspecified)
mutches [women's close-fitting caps]	each; dozen
myrrh [a yellow to reddish brown aromatic gum resin]	avoirdupois pound (453.592 g)
nails	thousand of 1,200 in number; half and whole barrel (quantity unspecified); sum of 10,000 avoirdupois pounds (4,535.900 kg)
napkins	dozen
nard [an ointment]	avoirdupois pound (453.592 g)
neats' tongues [bovine]	each
neckerchiefs	dozen
needles	thousand of 1,000 in number; twelve thousand
nigella [an herb]	avoirdupois pound (453.592 g)
nightcaps	dozen
Normandy canvas	hundred ells of 45 inches (1.143 m) each
nutmegs	avoirdupois pound (453.592 g); hundredweight of 108 avoirdupois pounds (48.988 kg)
nuts	barrel of 3 bushels (ca. 1.06 hl)
nux indica [a drug]	avoirdupois pound (453.592 g)
nux vomica [an alkaloid; a poison]	avoirdupois pound (453.592 g)
oakum [a heavy, twisted hemp or jute fiber used for caulking]	hundredweight of 112 avoirdupois pounds (50.802 kg)
oars	each; hundred of 120 in number

Appendix A

Imports (continued)

Description of Merchandise	Bulk Rated
oats	quarter (ca. 2.82 hl)
ocher [a red or yellow iron ore]	barrel (weight unspecified)
oil	gallon (3.785 l); barrel of 31½ gallons (ca. 1.19 hl); tun of 252 gallons (ca. 9.54 hl)
oilbanum [frankincense]	avoirdupois pound (453.592 g)
olives	hogshead (weight unspecified)
onions	barrel (weight unspecified); hundred bunches of 2,500 in number
onion seeds	hundredweight of 112 avoirdupois pounds (50.802 kg)
opium	avoirdupois pound (453.592 g)
opopanax [an odorous gum resin]	avoirdupois pound (453.592 g)
oranges and lemons	thousand of 1,000 in number
orchil [a violet dye]	hundredweight of 112 avoirdupois pounds (50.802 kg)
origanum [an herb]	avoirdupois pound (453.592 g)
orpiment [an orange to yellow mineral]	hundredweight of 112 avoirdupois pounds (50.802 kg)
orrice [orris; a flavoring material]	hundredweight of 112 avoirdupois pounds (50.802 kg)
osmund [a superior iron used for arrowheads, etc.]	last of 12 barrels (weight unspecified)
otter skins	each
ounce skins [of wildcats]	each
oxen	each
pack needles	thousand of 1,000 in number
packthread	hundredweight of 112 avoirdupois pounds (50.802 kg)
painted cloths	dozen
painted coffers	nest of 3 in number
painted papers	gross; ream of 20 quires or 500 in number
painted trenchers [platters or trays]	gross
painter's or linseed oil	barrel of 31½ gallons (ca. 1.19 hl)
pennels [pannels; saddle pads]	hundred of 100 in number
pans	dozen; hundredweight of 112 avoirdupois pounds (50.802 kg)
paper	hundred leaves; ream of 20 quires or 500 in number; bundle of 40 quires of 1,000 in number; bale or 10 reams or 200 quires or 5,000 in number
parchment	dozen; hundred of 100 in number
Paris mantles [cloaks]	each

Imports (continued)

Description of Merchandise	Bulk Rated
parmacety [spermaceti; an ointment]	avoirdupois pound (453.592 g)
partisans [long-shafted spears or halberds]	dozen
pasments (trimmings)	avoirdupois pound (453.592 g); bolt (dimensions unspecified)
pasteboard	hundredweight of 112 avoirdupois pounds (50.802 kg)
paving stones	each; thousand of 1,200 in number
pearling [a lace of silk or thread]	ell of 45 inches (1.143 m)
pearls	troy ounce (31.103 g)
pears or apples	barrel of 3 bushels (ca. 1.06 hl)
peas	quarter (ca. 2.82 hl)
penners [cases worn at the waist for holding pens]	gross
pepper	avoirdupois pound (453.592 g); hundredweight of 100 avoirdupois pounds (45.359 kg)
perry [fermented pear juice]	tun of 252 gallons (ca. 9.54 hl)
petticoats	each
pewter	hundredweight of 112 avoirdupois pounds (50.802 kg)
pheasants	dozen
pike heads	each
pikes	each
pillars	dozen; gross
pincushions	dozen
pine	avoirdupois pound (453.592 g)
pink root [an herb]	avoirdupois pound (453.592 g)
pins	twelve thousand of 14,400 in number
pinsons [pincers]	dozen
pipes	dozen; gross; bale of 10 gross or 1,440 in number
pipe staves	hundred of 120 in number
pitch or tar	last of 12 barrels (ca. 17.76 hl)
planes	dozen
plaster of Paris	mount of 3 thousandweight (1,524.060 kg)
plate [of precious metals; used for coating]	avoirdupois ounce (28.350 g)
plates	each; dozen; hundred of 100 in number; barrel of 300 in number
playing cards	gross
playing tables	pair; dozen

Appendix A

Imports (continued)

Description of Merchandise	*Bulk Rated*
plums	hundredweight of 100 avoirdupois pounds (45.359 kg)
points [usually of leather]	gross; great gross or 1,728 in number
poldavies [a coarse canvas or sacking for sails]	bolt (dimensions unspecified)
pomegranates	thousand of 1,000 in number
pork	ton of 2,240 avoirdupois pounds (1,016.040 kg)
potatoes	hundredweight of 112 avoirdupois pounds (50.802 kg)
pots	dozen; hundred of 100 in number; avoirdupois pound (453.592 g); hundredweight of 112 avoirdupois pounds (50.802 kg)
powder	hundredweight of 112 avoirdupois pounds (50.802 kg)
pouch rings	gross
primers (printed)	gross
prunes	hundredweight of 112 avoirdupois pounds (50.802 kg)
pulleys	dozen; gross
pumice [volcanic glass used for smoothing and polishing]	ton of 2,240 avoirdupois pounds (1,016.040 kg)
puppets (for children)	gross
purling wire	twelve avoirdupois pounds (5.443 kg)
purses	gross
quails	dozen
quassia [a tonic]	hundredweight of 112 avoirdupois pounds (50.802 kg)
quern stones [millstones]	last of 12 pairs
quicksilver	avoirdupois pound (453.592 g)
quills	thousand of 1,000 in number
quilts	each; dozen
quinces	hundred of 100 in number
quinine [an alkaloid]	avoirdupois ounce (28.350 g)
rackets	each; dozen
racks (for crossbows)	each
rafters	hundred of 100 in number
rags	ton of 2,240 avoirdupois pounds (1,016.040 kg)
raisins	hundredweight of 112 avoirdupois pounds (50.802 kg)
rape oil [a lubricant, illuminant, and food]	barrel of 31½ gallons (ca. 1.19 hl); last (capacity unspecified)

English Import and Export Items 125

Imports (continued)

Description of Merchandise	Bulk Rated
rapeseeds	quarter (capacity or weight unspecified)
rapiers [straight two-edged swords]	dozen
rattles (for children)	gross
razors	dicker of 10 in number; gross
recorders [flutes]	case of 5 in number
red earth (for painters)	hundredweight of 112 avoirdupois pounds (50.802 kg)
red hides	dicker of 10 in number
red lead	hundredweight of 112 avoirdupois pounds (50.802 kg)
reeds or canes	thousand of 1,200 in number
regals [small portable organs with reed pipes]	pair
rhapontic [a drug]	avoirdupois pound (453.592 g)
rhinehurst [the wood]	hundredweight of 112 avoirdupois pounds (50.802 kg)
rhubarb	hundredweight of 112 avoirdupois pounds (50.802 kg)
ribbons	avoirdupois pound (453.592 g); gross
rice	hundredweight of 112 avoirdupois pounds (50.802 kg)
rims (for sieves)	gross
rings	avoirdupois pound (453.592 g); gross; two gross
roan skins [generally implying "reddish brown"]	dozen
rods	bundle (quantity unspecified)
rosewater (in Venice glasses)	dozen
rosewood	ton of 2,240 avoirdupois pounds (1,016.040 kg)
rosin	hundredweight of 112 avoirdupois pounds (50.802 kg)
round boxes	dozen
rugs	each
rungs	hundred of 100 in number
sables [the pelts or skins]	timber of 40 in number
sackcloth [a coarse cloth for garments]	yard; ell of 45 inches (1.143 m); piece of 24 yards (ca. 21.95 m) in length and 7 quarters (ca. 1.60 m) in breadth; roll (dimensions unspecified)
saddles	each
saffron	avoirdupois pound (453.592 g)
sal ammoniac [ammonium chloride]	avoirdupois pound (453.592 g)

Appendix A

Imports (continued)

Description of Merchandise	Bulk Rated
salmon	barrel of 42 gallons (ca. 1.59 hl); last of 6 pipes or 504 gallons (ca. 19.08 hl)
salop [salep; a food and demulcent]	avoirdupois pound (453.592 g)
salt	bushel (35.238 l); barrel (capacity or weight unspecified); wey of 42 bushels (ca. 14.80 hl)
salt fish	last of 12 barrels (ca. 17.76 hl)
salt hides	dicker of 10 in number
saltpeter	hundredweight of 112 avoirdupois pounds (50.802 kg)
sandboxes	gross
sanguis draconis [an herb]	avoirdupois pound (453.592 g)
sarcenet [a soft thin silk]	yard; ell of 45 inches (1.143 m)
sarsaparilla	hundredweight of 100 avoirdupois pounds (45.359 kg)
sassafras [an herb]	hundredweight of 112 avoirdupois pounds (50.802 kg)
satin	yard; piece of 24 yards (ca. 21.95 m) in length and 7 quarters (ca. 1.60 m) in breadth
saws	each; dozen
says [fine woolen cloth]	piece of 24 yards (ca. 21.95 m) in length and 7 quarters (ca. 1.60 m) in breadth
scales (for scabbards)	thousand of 1,000 in number
scammony [a violent cathartic]	avoirdupois pound (453.592 g)
scissors	gross
scoops	dozen
sea holly [an herb, an aphrodisiac]	hundredweight of 112 avoirdupois pounds (50.802 kg)
sebesten [a demulcent]	avoirdupois pound (453.592 g)
seed pearl [a very small and often irregular pearl]	troy ounce (31.103 g)
sendal [a thin silk]	piece of 24 yards (ca. 21.95 m) in length and 7 quarters (ca. 1.60 m) in breadth
senna [a purgative]	avoirdupois pound (453.592 g)
setwall [garden heliotrope]	hundredweight of 112 avoirdupois pounds (50.802 kg)
shears	pair; dozen
sheep	score
sheets	dozen
shirts	each; dozen
shumack or blacking [sumac; for tanning and dyeing]	hundredweight of 112 avoirdupois pounds (50.802 kg)

English Import and Export Items

Imports (continued)

Description of Merchandise	Bulk Rated
shuttles [for weaving]	dozen
silk	pound; yard; ell of 45 inches (1.143 m); piece of 24 yards (ca. 21.95 m) in length and 7 quarters (ca. 1.60 m) in breadth
silver	mast of 2½ avoirdupois pounds (1.134 kg)
skins	each; dicker of 10 in number; dozen; timber of 40 in number; hundred of 100 in number
sleeves	pair
slip [a mixture of fine clay and water used in casting]	barrel (weight unspecified)
smalt [a powder used as a colorant for glass]	avoirdupois pound (453.592 g)
snuff [pulverized tobacco]	avoirdupois pound (453.592 g)
snuffers [for candles]	dozen
soap	hundredweight of 112 avoirdupois pounds (50.802 kg); barrel of 32 gallons (ca. 1.48 hl); last of 12 barrels (ca. 17.76 hl)
spangles [small pieces of shiny metal used for ornamentation]	thousand of 1,000 in number
Spanish skins	dozen
spars	each; hundred of 120 in number
spears	hundred of 100 in number
spectacle cases	gross
spectacles	gross
spicknel [an herb]	hundredweight of 112 avoirdupois pounds (50.802 kg)
spodium [a powder; soot from melting metals or vegetable ash]	avoirdupois pound (453.592 g)
sponges	avoirdupois pound (453.592 g); dozen
spoons	gross
sprigs [brads]	sum of 10,000 avoirdupois pounds (4,535.900 kg)
spruce iron [from Prussia]	hundredweight of 112 avoirdupois pounds (50.802 kg)
spurs	pair
squinancy [an herb]	avoirdupois pound (453.592 g)
squirts [syringes]	dozen
standishes [inkstands]	each; dozen
starch	hundredweight of 112 avoirdupois pounds (50.802 kg)
stavesacre [a violent emetic and cathartic]	hundredweight of 112 avoirdupois pounds (50.802 kg)

Imports (continued)

Description of Merchandise	Bulk Rated
steel	hundredweight of 112 avoirdupois pounds (50.802 kg); half-barrel (weight unspecified)
stirrups	dozen pairs
stitched cloth	ell of 45 inches (1.143 m)
stockfish [fish dried hard in the open air without salt]	hundred of 120 in number; last of 12 barrels (ca. 17.76 hl)
stonebirds (whistles)	gross
stones	each; last (quantity or weight unspecified); hundred of 120 in number; thousand of 1,200 in number; avoirdupois pound (453.592 g); ton of 2,240 avoirdupois pounds (1,016.040 kg)
stools	dozen
storax [a resin]	avoirdupois pound (453.592 g); hundredweight of 112 avoirdupois pounds (50.802 kg)
storks	dozen
strings	hundred of 100 in number; gross
stuffs (mixed with wool)	yard
sturgeon	firkin of 8 ale gallons (ca. 3.70 dkl); barrel of 32 ale gallons (ca. 1.48 hl)
succade [a preserve or confection made from fruit]	avoirdupois pound (453.592 g)
sugar	hundredweight of 108 avoirdupois pounds (48.988 kg); chest of 3 hundredweight (146.964 kg)
sugar candy	half-chest (weight unspecified)
sumac	hundredweight of 112 avoirdupois pounds (50.802 kg)
swan quills	thousand of 1,000 in number
swans	each
swan skins	each
sword blades	dozen
swords	each
syrup	avoirdupois pound (453.592 g)
table books	dozen
tables	pair
tacks [all types]	thousand of 1,000 in number
taffeta [a delicate fabric with a lustrous surface woven of various fibers]	yard

English Import and Export Items 129

Imports (continued)

Description of Merchandise	Bulk Rated
tallow [animal fats used for candles]	barrel (weight unspecified); hundredweight of 112 avoirdupois pounds (50.802 kg)
tamarinds [a preservative, laxative, and dye]	avoirdupois pound (453.592 g)
tankards	flock of 40 in number
tapestry	ell of 45 inches (1.143 m)
tapioca	hundredweight of 112 avoirdupois pounds (50.802 kg)
tar	barrel of 31½ gallons (ca. 1.19 hl); last of 12 barrels (ca. 14.28 hl)
targets	each
tarras [trass; a light-colored volcanic tuff]	barrel or bushel (capacity or weight unspecified)
tassels	thousand of 1,000 in number; pipe (quantity unspecified)
teasels [for raising a nap on woolen cloth]	thousand of 1,000 in number
thimbles	thousand of 1,000 in number
thread	avoirdupois pound (453.592 g); twelve avoirdupois pounds (5.443 kg); hundredweight of 112 avoirdupois pounds (50.802 kg); bale of 100 bolts; butt (weight or quantity unspecified)
thrums [fringe of warp threads left on the loom after weaving]	avoirdupois pound (453.592 g)
ticking [a linen or cotton fabric used for upholstering]	ell of 45 inches (1.143 m)
ticks	each; dozen
tiles	thousand of 1,200 in number
tin	hundredweight of 112 avoirdupois pounds (50.802 kg)
tinfoil	gross
tin glass	hundredweight of 112 avoirdupois pounds (50.802 kg)
tinsel	yard; ell of 45 inches (1.143 m)
tips (for horns)	thousand of 1,000 in number
tobacco	avoirdupois pound (453.592 g)
tongs	dozen
tools (carving)	gross
toothpicks	gross
touchboxes [carried lighted tinder for firing matchlocks]	dozen

Appendix A

Imports (continued)

Description of Merchandise	Bulk Rated
tow [a rope, heavy twine, or chain used for towing]	hundredweight of 112 avoirdupois pounds (50.802 kg)
trane [probably train oil, (whale oil)]	barrel (weight unspecified)
trass [a light-colored volcanic tuff]	barrel (weight unspecified)
trays	each; shock of 60 in number
treacle [a medicinal compound used against poisons]	avoirdupois pound (453.592 g); barrel (weight unspecified)
trenchers [platters or trays]	gross; thousand of 1,000 in number
truffles [a fungus]	avoirdupois pound (453.592 g)
trunks	dozen
tucks [rapiers]	each
turbith [the meaning is not certain]	avoirdupois pound (453.592 g)
turpentine	hundredweight of 112 avoirdupois pounds (50.802 kg)
twine	avoirdupois pound (453.592 g)
twist [the meaning is not certain]	dozen
valances	each
valonia [dried acorn cups used in tanning or dressing leather]	hundredweight of 112 avoirdupois pounds (50.802 kg)
varnish	hundredweight of 112 avoirdupois pounds (50.802 kg)
vellum [calfskin used for manuscripts]	each
velvet	yard
Venice gold or silver	troy pound (373.242 g)
Venice purses	dozen
Venice ribbons	dozen avoirdupois pounds (5.443 kg)
Venice turpentine	avoirdupois pound (453.592 g)
verdigris [a green or greenish blue pigment used in paints and medicine]	hundredweight of 100 avoirdupois pounds (45.359 kg)
verditer [a blue or green pigment used in paints]	hundredweight of 112 avoirdupois pounds (50.802 kg)
verjuice [sour juice of green apples, etc.]	avoirdupois pound (453.592 g)
vermilion [a red pigment used for artists' color]	hundredweight of 112 avoirdupois pounds (50.802 kg)
vinegar	tun of 252 gallons (ca. 9.54 hl)
viols [bowed stringed instruments]	each
virginals [small spinets]	pair
vises	dozen

Imports (continued)

Description of Merchandise	Bulk Rated
vitex [agnus castus; a stimulant drug]	avoirdupois pound (453.592 g)
vitry [a light durable canvas]	balet (quantity unspecified); bolt (quantity unspecified)
vizards [masks]	dozen
wadmol [a coarse woolen fabric]	yard; ell of 45 inches (1.143 m)
wafers	avoirdupois pound (453.592 g)
wainscots	hundred of 120 in number
walkers earth [for finishing cloth]	hundredweight of 112 avoirdupois pounds (50.802 kg)
warming pans or bedpans	dozen
wax	avoirdupois pound (453.592 g); hundredweight of 108 avoirdupois pounds (48.988 kg)
weld [a yellow dye]	hundredweight of 112 avoirdupois pounds (50.802 kg)
whale fins	tun (weight unspecified)
wheat	quarter (ca. 2.82 hl)
whetstones [for whetting edge tools]	hundred of 120 in number
whipcord [a thin tough cord made of braided or twisted hemp or catgut]	avoirdupois pound (453.592 g); shock of 60 in number
whistles	gross
white or red lead	hundredweight of 112 avoirdupois pounds (50.802 kg)
whitings [fish]	last (weight unspecified)
wimples [women's cloth head and neck coverings]	each
wine	tun of 252 gallons (ca. 9.54 hl)
wire (for clavichords)	avoirdupois pound (453.592 g)
woad [a blue dye]	hundredweight of 112 avoirdupois pounds (50.802 kg); balet (weight unspecified); poke (weight unspecified); ton of 2,240 avoirdupois pounds (1,016.040 kg)
wolf skins	each
wood	thousand of 1,000 in number; hundredweight of 112 avoirdupois pounds (50.802 kg)
wool	avoirdupois pound (453.592 g); hundredweight of 112 avoirdupois pounds (50.802 kg)
wool cards [for carding wool]	dozen

Appendix A

Imports (continued)

Description of Merchandise	Bulk Rated
wormseed [an anthelmintic]	avoirdupois pound (453.592 g); hundredweight of 112 avoirdupois pounds (50.802 kg)
worsted [a closely woven fabric made from woolen yarns]	piece of 24 yards (ca. 21.95 m) in length and 7 quarters (ca. 1.60 m) in breadth
worsted yarn	twelve avoirdupois pounds (5.443 kg)
wrests (for virginals) [keys or wrenches used for turning wrest pins]	gross
writing tables	dozen
yarn	avoirdupois pound (453.592 g); twelve avoirdupois pounds (5.443 kg); hundredweight of 112 avoirdupois pounds (50.802 kg)
zaffre [an impure cobalt oxide powder used in manufacture of smalt]	avoirdupois pound (453.592 g)
zedoaria [zedoary; a stimulant]	avoirdupois pound (453.592 g)

EXPORTS

Description of Merchandise	Bulk Rated
alabaster [a compact variety of fine-textured gypsum]	load (weight unspecified)
alum [an emetic or astringent]	hundredweight of 108 avoirdupois pounds (48.988 kg)
anvils	hundredweight of 112 avoirdupois pounds (50.802 kg)
apothecary and confectionary wares	hundredweight of 112 avoirdupois pounds (50.802 kg)
apples	bushel (35.238 l)
aqua vitae [a strong liquor; brandy, whiskey]	hogshead (capacity or weight unspecified)
ashes	barrel (ca. 1.48 hl); last of 12 barrels (ca. 17.76 hl)
bacon	flitch (a side of cured hog meat)
bags	dozen
baize [a coarsely woven woolen or cotton fabric]	piece of 24 yards (ca. 21.95 m) in length and 7 quarters (ca. 1.60 m) in breadth
bandoliers	hundred of 100 in number
barrel staves	thousand of 1,200 in number
beef	barrel of 32 wine gallons (ca. 1.21 hl)
beer	tun of 252 gallons (ca. 9.54 hl)

English Import and Export Items

Exports (continued)

Description of Merchandise	Bulk Rated
beeswax	hundredweight of 108 avoirdupois pounds (48.988 kg)
bell metal	hundredweight of 112 avoirdupois pounds (50.802 kg)
bellows	dozen
billets [short chunky pieces of firewood]	thousand of 1,000 in number
birdlime [an adhesive substance]	hundredweight of 112 avoirdupois pounds (50.802 kg)
bones (ox)	thousand of 1,000 in number
bonnets	hundred of 100 in number
books	hundredweight of 112 avoirdupois pounds (50.802 kg)
boots	thirty pairs
brass and brass wares	hundredweight of 100 avoirdupois pounds (45.359 kg); barrel (weight unspecified)
bricks	thousand of 1,200 in number
bridle bits	gross
bridles	dozen
brushes	dozen
bullets	thousand of 1,000 in number
butter	barrel of 32 ale gallons (ca. 1.48 hl)
buttons	gross
cable yarn [for ropes]	stone of 14 avoirdupois pounds (6.350 kg)
calf skins	dozen
cambodium [a drug]	avoirdupois pound (453.592 g)
candles	twelve avoirdupois pounds (5.443 kg); barrel (weight unspecified)
caps	dozen
cardboards	gross
cards	dozen; hundredweight of 112 avoirdupois pounds (50.802 kg)
carpets	each
carts	gross
cheese	hundredweight of 112 avoirdupois pounds (50.802 kg); wey (weight unspecified)
cloth (Scottish woolen)	ell of 45 inches (1.143 m)
coaches or chariots	each
coal	chalder of 2,000 avoirdupois pounds (907.180 kg) before 1676; 2,240 avoirdupois pounds (1,016.040 kg) afterward
codlings [young cod]	last of 496 in number
coney skins [rabbit]	hundred of 100 in number

Appendix A

Exports (continued)

Description of Merchandise	Bulk Rated
copper and copper wares	avoirdupois pound (453.592 g); hundredweight of 112 avoirdupois pounds (50.802 kg)
copperas [a green dye]	hundredweight of 112 avoirdupois pounds (50.802 kg)
corbels [architectural supports]	dozen
cordage [ship cords or ropes]	hundredweight of 112 avoirdupois pounds (50.802 kg)
corn [grain, generally wheat]	quarter (ca. 2.82 hl)
cottons	hundred goads of 4½ feet (1.371 m) each
coverlets	each
culm [refuse coal]	chalder (weight unspecified)
cushions	dozen
deals [boards of fir or pine]	hundred of 120 in number
dimity [a sheer cotton fabric]	yard
dornick [a coarse damask of wool and silk]	yard
dozens [coarse woolen cloth]	piece of 24 yards (ca. 21.95 m) in length and 7 quarters (ca. 1.60 m) in breadth
eggs	barrel (quantity unspecified)
emery stones	hundredweight of 112 avoirdupois pounds (50.802 kg)
feathers (for beds)	hundredweight of 112 avoirdupois pounds (50.802 kg)
figuretto [figured cloth]	piece of 24 yards (ca. 21.95 m) in length and 7 quarters (ca. 1.60 m) in breadth
fitches [the fur or pelt of the fitch]	timber of 40 in number
flannel	yard
flax [the bast fiber]	hundredweight of 112 avoirdupois pounds (50.802 kg)
flocks [woolen or cotton refuse]	hundredweight of 112 avoirdupois pounds (50.802 kg)
frieze [a coarse woolen cloth]	yard
fustians [a strong cotton and linen fabric]	yard
galls [the meaning is not certain]	hundredweight of 112 avoirdupois pounds (50.802 kg)
garters	gross
girdles	gross
glass	hundredweight of 112 avoirdupois pounds (50.802 kg); barrel (weight unspecified); chest (weight or quantity unspecified)

English Import and Export Items 135

Exports (continued)

Description of Merchandise	Bulk Rated
glovers' clippings	maund of 2 or 3 pecks (ca. 1.76 or ca. 2.64 dkl); fatt (weight unspecified)
gloves	dozen; gross
glue	hundredweight of 112 avoirdupois pounds (50.802 kg)
goats	each
goose quills	thousand of 1,000 in number
grindstones [millstones]	chalder (varied in number according to size)
grograms [a coarse, loosely woven fabric of silk and wool]	piece of 24 yards (ca. 21.95 m) in length and 7 quarters (ca. 1.60 m) in breadth
gunpowder	hundredweight of 100 avoirdupois pounds (45.359 kg)
guts (ox)	barrel (weight unspecified)
haberdashery	hundredweight of 112 avoirdupois pounds (50.802 kg)
hair	hundredweight of 112 avoirdupois pounds (50.802 kg)
hakes [fish]	hundred of 100 in number
harnesses	pair
hats	dozen
hawks' hoods	dozen
hemp	stone of 16 avoirdupois pounds (7.257 kg)
hempseed	quarter (ca. 2.82 hl)
herrings	cade of 500 to 1,000 in number; barrel of 30 gallons; last of 12,000 in number
hides	dicker of 10 in number
holsters	dozen pairs
honey	barrel of 31½ gallons (ca. 1.19 hl)
hooks	gross
hoops (for barrels)	thousand of 1,000 in number
hops	hundredweight of 112 avoirdupois pounds (50.802 kg)
horns	hundred of 100 in number; thousand of 1,000 in number
horse collars	hundred of 100 in number
horses	each
horse tails	hundred of 100 in number
hose	pair; dozen pairs
Irish mantles [cloaks]	each
iron	stone of 14 avoirdupois pounds (6.350 kg); hundredweight of 112 avoirdupois pounds (50.802 kg); ton of 2,240 avoirdupois pounds (1,016.040 kg)

Appendix A

Exports (continued)

Description of Merchandise	Bulk Rated
iron pots	dozen
Jedburgh staves	hundred of 120 in number
kerseys [coarse ribbed woolen cloth for hose and work clothes]	piece of 16 to 18 yards (ca. 14.63 to ca. 16.40 m) in length and 4 quarters (ca. 0.91 m) in breadth
kettles	hundredweight of 112 avoirdupois pounds (50.802 kg)
kid skins	hundred of 100 in number
knives	twenty dozen
lace (gold and silver)	pound
lamb skins	hundred of 120 in number
lampreys [eels]	thousand of 1,000 in number
lead	barrel (weight unspecified); fother of 2,100 avoirdupois pounds (952.539 kg)
leather and leather wares	avoirdupois pound (453.592 g); dicker of 10 in number; hundred of 100 in number; gross
lime	chalder (weight unspecified)
linen	forty ells of 45 inches (1.143 m) each
lings [deep-sea fish]	hundred of 124 in number; last of 12 barrels (ca. 17.76 hl)
linseed	quarter (ca. 2.82 hl)
lint [flax]	stone of 14 avoirdupois pounds (6.350 kg)
longcloth [a fine bleached cotton cloth]	piece of 24 yards (ca. 21.95 m) in length and 7 quarters (ca. 1.60 m) in breadth
madder [a dye]	poke (weight unspecified)
malt	quarter (ca. 2.82 hl)
maps and charts	hundredweight of 112 avoirdupois pounds (50.802 kg)
masts	dozen
meal [various grains]	quarter (ca. 2.82 hl); last of 12 barrels (ca. 17.76 hl)
mittens	thousand pairs
mustard seed	hundredweight of 112 avoirdupois pounds (50.802 kg)
nails	hundredweight of 112 avoirdupois pounds (50.802 kg); sum of 10,000 avoirdupois pounds (4,535.900 kg)
needles	twenty gross
nightcaps	thirty dozen
nuts	barrel of 3 bushels (ca. 1.06 hl); last (weight unspecified)

English Import and Export Items 137

Exports (continued)

Description of Merchandise	Bulk Rated
oatmeal	bushel (35.238 l); barrel of 2 hundredweight (101.604 kg)
ocher [a red or yellow iron ore]	hundredweight of 112 avoirdupois pounds (50.802 kg)
oil	barrel of 31½ gallons (ca. 1.19 hl); tun of 252 gallons (ca. 9.54 hl)
otter skins	each
oxen	each
oysters (pickled)	barrel of 32 gallons (ca. 1.48 hl)
painted books	fatt of 4 bales
pans (of brass)	hundredweight of 112 avoirdupois pounds (50.802 kg)
paper	avoirdupois pound (453.592 g); hundredweight of 112 avoirdupois pounds (50.802 kg)
parchment	roll of 60 skins
pasteboards	gross
pennystones [a cloth]	piece of 12 or 13 yards (ca. 10.97 or ca. 11.89 m) in length and 3½ to 6½ quarters (ca. 0.80 to ca. 1.49 m) in breadth
pewter	hundredweight of 112 avoirdupois pounds (50.802 kg)
pictures	hundredweight of 112 avoirdupois pounds (50.802 kg)
pilchards [sardines]	ton of 2,240 avoirdupois pounds (1,016.040 kg)
pipe staves	hundred of 120 in number
pistols	pair
pitch	barrel (ca. 1.48 hl)
plaiding [plaid cloth]	ell of 45 inches (1.143 m)
points (of leather)	gross
pork	barrel (weight unspecified)
pots (of brass)	hundredweight of 112 avoirdupois pounds (50.802 kg)
powder	barrel (weight unspecified)
purses	gross
rape cakes	thousand of 1,000 in number
rapeseeds	quarter (capacity or weight unspecified)
red earth [dye for painting]	hundredweight of 112 avoirdupois pounds (50.802 kg)
red heath [the plant; heather]	hundredweight of 112 avoirdupois pounds (50.802 kg)

Appendix A

Exports (continued)

Description of Merchandise	Bulk Rated
red ocher [an iron ore]	hogshead (weight unspecified)
ribbons	avoirdupois pound (453.592 g); gross
ropes	hundredweight of 112 avoirdupois pounds (50.802 kg)
rugs	yard
sackcloth [a coarse cloth for garments]	ell of 45 inches (1.143 m)
saddles	each
saffron	avoirdupois pound (453.592 g)
sail canvas	ell of 45 inches (1.143 m)
salmon	last of 6 pipes or 504 gallons (ca. 19.08 hl)
salt	chalder (weight unspecified); barrel (weight unspecified)
saltpeter	hundredweight of 112 avoirdupois pounds (50.802 kg)
says [fine woolen cloth]	piece of 24 yards (ca. 21.95 m) in length and 7 quarters (ca. 1.60 m) in breadth
scarlets [rich, brightly colored textiles]	yard
serges [durable twilled fabrics with pronounced diagonal ribbing]	ell of 45 inches (1.143 m)
sheep skins	hundred of 100 in number
shoes	hundred pairs; hundred dozen pairs
shovels	each
silk	avoirdupois pound (453.592 g)
skins	each; dicker of 10 in number; dozen; hundred of 100 in number; thousand of 1,000 in number
soap	barrel of 32 gallons (ca. 1.48 hl)
spars	hundred of 120 in number
sprats [herrings, pilchards]	cade of 500 to 1,000 in number; last of 12,000 in number
starch	hundredweight of 112 avoirdupois pounds (50.802 kg)
staves (for barrels)	thousand of 1,200 in number
steel	hundredweight of 112 avoirdupois pounds (50.802 kg)
stockings	dozen
stones (slate)	thousand of 1,200 in number
stuffs (woolen)	avoirdupois pound (453.592 g)
sugar	hundredweight of 108 avoirdupois pounds (48.988 kg)

Exports (continued)

Description of Merchandise	Bulk Rated
swine	each
sword belts	gross
sword blades	score
swords	each
tackle [a ship's rigging]	stone of 14 avoirdupois pounds (6.350 kg)
tallow [animal fats used for candles]	hundredweight of 112 avoirdupois pounds (50.802 kg); barrel (weight unspecified)
tapestry	avoirdupois pound (453.592 g)
tar	barrel of 31½ gallons (ca. 1.19 hl)
tin	hundredweight of 112 avoirdupois pounds (50.802 kg)
tips of horns	thousand of 1,000 in number
tobacco pipes	gross
velures [velvet fabric]	piece of 24 yards (ca. 21.95 m) in length and 7 quarters (ca. 1.60 m) in breadth
vinegar	tun of 252 gallons (ca. 9.54 hl)
virginals [small spinets]	pair
watches	each
wax	avoirdupois pound (453.592 g); hundredweight of 108 avoirdupois pounds (48.988 kg)
whale fins	gross
wheat	quarter (ca. 2.82 hl)
wine	butt of 126 gallons (ca. 4.77 hl); tun of 252 gallons (ca. 9.54 hl)
woad [a blue dye]	ton of 2,240 avoirdupois pounds (1,016.040 kg)
wood	hundredweight of 112 avoirdupois pounds (50.802 kg); ton of 2,240 avoirdupois pounds (1,016.040 kg)
wool	stone of 14 avoirdupois pounds (6.350 kg)
worsted [a closely woven fabric made from woolen yarns]	piece of 24 yards (ca. 21.95 m) in length and 7 quarters (ca. 1.60 m) in breadth
yarn	avoirdupois pound (453.592 g)

Appendix B

British Pre-Imperial Units

These tables show the dimensions and the intra-unit proportions of standard weights and measures units in the British Isles before the passage of the Imperial Weights and Measures Act of 1824. There are no local variations recorded in any of the tables and, unless otherwise indicated, the value given for each unit remained constant following its initial description in a statute or other government enactment.

Relationship of English Linear Measures to Scots, Irish, and Welsh Linear Measures

1 Scots inch	=	1.0054054 English inches
1 Irish inch	=	1 English inch
1 Welsh inch	=	1 English inch
1 Scots foot	=	12.064864 English inches
1 Irish foot	=	12 English inches
1 Welsh foot	=	12 English inches
1 Scots mile	=	5,952 English feet
1 Irish mile	=	6,720 English feet
1 Welsh mile	=	5,280 English feet

English Linear Measures

inches	palms	Gunter's links	spans	feet	cubits	yards	paces	fathoms	poles or perches	Gunter's chains	furlongs	miles
1												
3	1											
7.92	2.64	1										
9	3	1.136	1									
12	4	1.515	1.33	1								
18	6	2.272	2	1.5	1							
36	12	4.545	4	3	2	1						
60	20	7.575	6.67	5	3.33	1.67	1					
72	24	9.091	8	6	4	2	1.2	1				
198	66	25	22	16.5	11	5.5	3.30	2.75	1			
792	264	100	88	66	44	22	13.2	11	4	1		
7,920	2,640	1,000	880	660	440	220	132	110	40	10	1	
63,360	21,120	8,000	7,040	5,280	3,520	1,760	1,056	880	320	80	8	1

British Pre-Imperial Units

Scots Linear Measures[a]

English inches	Gunter's links	English feet	ells	falls	Gunter's chains	furlongs	miles
1							
8.928	1						
12	1.344	1					
37.2	4.166	3.1	1				
223.2	25	18.6	6	1			
892.8	100	74.4	24	4	1		
8,928	1,000	744	240	40	10	1	
71,424	8,000	5,952	1,920	320	80	8	1

[a] Unless otherwise specified, the units are Scots.

Note the following intra-unit proportions:

	Scots inches	Scots feet	Scots ells
	1		
	12	1	
	37	3.083	1

Irish Linear Measures

inches	palms	Gunter's links	spans	feet	cubits	yards	paces	fathoms	poles or perches	Gunter's chains	furlongs	miles
1												
3	1											
10.08	3.36	1										
9	3	0.893	1									
12	4	1.190	1.33	1								
18	6	1.785	2	1.5	1							
36	12	3.571	4	3	2	1						
60	20	5.952	6.67	5	3.33	1.67	1					
72	24	7.143	8	6	4	2	1.2	1				
252	84	25	28	21	14	7	4.2	3.5	1			
1,008	336	100	112	84	56	28	16.8	14	4	1		
10,080	3,360	1,000	1,120	840	560	280	168	140	40	10	1	
80,640	26,880	8,000	8,960	6,720	4,480	2,240	1,344	1,120	320	80	8	1

English Area Measures

square inches	square Gunter's links	square feet	square yards	square paces	square poles or perches	square Gunter's chains	roods	acres
1								
62.726	1							
144	2.295	1						
1,296	20.661	9	1					
3,600	57.392	25	2.778	1				
39,204	625	272.25	30.25	10.89	1			
627,264	10,000	4,356	484	174.24	16	1		
1,568,160	25,000	10,890	1,210	435.6	40	2.5	1	
6,272,640	100,000	43,560	4,840	1,742.4	160	10	4	1

145

Appendix B

Scots Area Measures[a]

square English inches	square Gunter's links	square English feet	square ells	square falls	square Gunter's chains	roods	acres
1							
79.709	1						
144	1.806	1					
1,383.8	17.361	9.61	1				
49,818.2	625	345.96	36	1			
797,091.8	10,000	5,535.36	576	16	1		
1,992,729.6	25,000	13,838.4	1,440	40	2.5	1	
7,970,918.4	100,000	55,353.6	5,760	160	10	4	1

[a] Unless otherwise specified, the units are Scots.

Irish Area Measures

square inches	square Gunter's links	square feet	square yards	square paces	square poles or perches	square Gunter's chains	roods	acres
1								
101.606	1							
144	1.417	1						
1,296	12.775	9	1					
3,600	35.431	25	2.778	1				
63,504	625	441	49	17.69	1			
1,016,064	10,000	7,056	784	283.04	16	1		
2,540,160	25,000	17,640	1,960	707.6	40	2.5	1	
10,160,640	100,000	70,560	7,840	2,830.4	160	10	4	1

Welsh Area Measures

square yards	erws	tyddyns	rhandirs	gavaels	trevs	maenols	cymwds	cantrevs
1								
4,320	1							
17,280	4	1						
69,120	16	4	1					
276,480	64	16	4	1				
1,105,920	256	64	16	4	1			
4,423,680	1,024	256	64	16	4	1		
55,296,000	12,800	3,200	800	200	50	12.5	1	
110,592,000	25,600	6,400	1,600	400	100	25	2	1

English Capacity Measures for Wine

cubic inches	pints	quarts	gallons	rundlets	barrels	tierces	hogs-heads	pun-cheons	butts	tuns
1										
28.875	1									
57.75	2	1								
231	8	4	1							
4,158	144	72	18	1						
7,276.5	252	126	31.5	1.75	1					
9,702	336	168	42	2.33	1.33	1				
14,553	504	252	63	3.5	2	1.5	1			
19,279	672	336	84	4.667	2.667	2	1.33	1		
29,106	1,008	504	126	7	4	3	2	1.5	1	
58,212	2,016	1,008	252	14	8	6	4	3	2	1

Appendix B

English Capacity Measures for Malted Beverages

cubic inches	pints	quarts	gallons	barrels	hogsheads
Ale (before 1688)					
1					
35.25	1				
70.5	2	1			
282	8	4	1		
9,024	256	128	32	1	
13,536	384	192	48	1.5	1
Ale and Beer (1688–1803)					
1					
35.25	1				
70.5	2	1			
282	8	4	1		
9,588	272	136	34	1	
14,382	408	204	51	1.5	1
Beer (before 1688) and Ale and Beer (1803–24)					
1					
35.25	1				
70.5	2	1			
282	8	4	1		
10,152	288	144	36	1	
15,228	432	216	54	1.5	1

English Cloth Measures

inches	nails	yards	ells
1			
2.25	1		
36	16	1	
45	20	1.25	1

English Coal Measures

avoirdupois pounds	bushels	chalders	keels
\multicolumn{4}{c}{1421–1676}			
1			
62.5	1		
2,000	32	1	
40,000	640	20	1
\multicolumn{4}{c}{1676–1824}			
1			
62.2	1		
2,240	36	1	
35,840	576	16	1

Scots Liquid Capacity Measures[a]

cubic English inches	gills	mutchkins	choppins	pints	quarts	gallons	barrels
1							
6.46275	1						
25.851	4	1					
51.702	8	2	1				
103.404	16	4	2	1			
206.808	32	8	4	2	1		
827.232	128	32	16	8	4	1	
6,617.856	1,024	256	128	64	32	8	1

[a] Unless otherwise specified, the units are Scots.

Irish Liquid Capacity Measures

cubic inches	noggins	pints	quarts	pottles	gallons	rundlets	barrels	tierces	hogs-heads	pun-cheons	pipes	tuns
1												
6.8	1											
27.2	4	1										
54.4	8	2	1									
108.8	16	4	2	1								
217.6	32	8	4	2	1							
3,916.8	576	144	72	36	18	1						
6,854.4	1,008	252	126	63	31.5	1.75	1					
9,139.2	1,344	336	168	84	42	2.33	1.33	1				
13,708.8	2,016	504	252	126	63	3.5	2	1.5	1			
18,278.4	2,688	672	336	168	84	4.667	2.667	2	1.33	1		
27,417.6	4,032	1,008	504	252	126	7	4	3	2	1.5	1	
54,835.2	8,064	2,016	1,008	504	252	14	8	6	4	3	2	1

Note: Irish ale measures are as follows: 10 gallons = 1 firkin; 2 firkins = 1 kilderkin; 2 kilderkins = 1 barrel; 8 barrels = 1 tun.

English Dry Capacity Measures

cubic inches	pints	quarts	gallons	pecks	bushels	quarters
1						
33.6	1					
67.2	2	1				
268.8	8	4	1			
537.6	16	8	2	1		
2,150.42	64	32	8	4	1	
17,203.36	512	256	64	32	8	1

Scots Dry Capacity Measures for Wheat, Peas, Beans, Rye, and White Salt[a]

cubic English inches	pints	lippies	pecks	firlots	bolls	chalders
1						
103.404	1					
137.333	1.3281	1				
549.333	5.3125	4	1			
2,197.335	21.25	16	4	1		
8,789.34	85	64	16	4	1	
140,629.44	1,360	1,024	256	64	16	1

[a] Unless otherwise specified, the units are Scots.

Scots Dry Capacity Measures for Oats, Barley, and Malt[a]

cubic English inches	pints	lippies	pecks	firlots	bolls	chalders
1						
103.404	1					
200.345	1.9375	1				
801.381	7.75	4	1			
3,205.524	31	16	4	1		
12,822.096	124	64	16	4	1	
205,153.53	1,984	1,024	256	64	16	1

[a] Unless otherwise specified, the units are Scots.

Irish Dry Capacity Measures

cubic inches	pints	quarts	pottles	gallons	pecks	bushels	strikes	barrels	quarters
1									
27.2	1								
54.4	2	1							
108.8	4	2	1						
217.6	8	4	2	1					
435.2	16	8	4	2	1				
1,740.8	64	32	16	8	4	1			
3,481.6	128	64	32	16	8	2	1		
6,963.2	256	128	64	32	16	4	2	1	
13,926.4	512	256	128	64	32	8	4	2	1

British Pre-Imperial Units

English Tower Weights

troy grains	pennyweights	ounces	pounds
1			
22.5	1		
450	20	1	
5,400	240	12	1

English Mercantile Weights

troy grains	pennyweights	ounces	pounds
1			
22.5	1		
450	20	1	
6,750	300	15	1

English Troy Weights

troy grains	pennyweights	ounces	pounds
1			
24	1		
480	20	1	
5,760	240	12	1

English Apothecary Weights

troy grains	scruples	drams	ounces	pounds
1				
20	1			
60	3	1		
480	24	8	1	
5,760	288	96	12	1

Appendix B

English Avoirdupois Weights

troy grains	drams	ounces	pounds	stone	hundred-weight	tons
1						
27.344	1					
437.5	16	1				
7,000	256	16	1			
98,000	3,584	224	14	1		
784,000	28,672	1,792	112	8	1	
15,680,000	573,440	35,840	2,240	160	20	1

English Lead Weights

avoirdupois pounds	stone	fotmals	weys	fothers
1				
12.5	1			
70	5.6	1		
175	14	2.5	1	
2,100	168	30	12	1

English Hay Weights

avoirdupois pounds	trusses	loads
Old		
1		
56	1	
2,016	36	1
New		
1		
60	1	
2,160	36	1

English Wool Weights

avoirdupois pounds	cloves or nails	stone	tods	weys	sacks	sarplers	lasts
1							
7	1						
14	2	1					
28	4	2	1				
182	26	13	6.5	1			
364	52	26	13	2	1		
728	104	52	26	4	2	1	
4,368	624	312	156	24	12	6	1

English Imaginary Mint Weights

blanks	perits	droits	mites	troy grains
1				
24	1			
480	20	1		
11,520	480	24	1	
230,400	9,600	480	20	1

Scots Troy Weights for Gold and Silver[a]

English troy grains	drops	ounces	pounds
1			
30	1		
480	16	1	
5,760	192	12	1

[a] Unless otherwise specified, the units are Scots.

Appendix B

Scots Troy Weights for Meal, Meat, Hemp, Unwrought Pewter, Flax, Lead, Iron, and Baltic and Dutch Goods[a]

English troy grains	drops	ounces	pounds	stone
1				
29.75	1			
476	16	1		
7,616	256	16	1	
121,856	4,096	256	16	1

[a] Unless otherwise specified, the units are Scots.

Scots Tron Weights[a]

English troy grains	drops	ounces	pounds	stone
1				
29.75	1			
476	16	1		
9,520	320	20	1	
152,320	5,120	320	16	1

[a] Unless otherwise specified, the units are Scots.

English, Scots, and Irish Quantity Measures

each	pairs	nests	dickers	dozen	score	flock or timber	shocks	hundreds	great hundreds	gross	thou-sands	great gross
1												
2	1											
3	1.5	1										
10	5	3.33	1									
12	6	4	1.2	1								
20	10	6.667	2	1.667	1							
40	20	13.33	4	3.33	2							
60	30	20	6	5	3	1.5						
100	50	33.33	10	8.33	5	2.5	1.667	1				
120	60	40	12	10	6	3	2	1.2	1			
144	72	46	14.4	12	7.2	3.6	2.4	1.44	1.2	1		
1,000	500	333.33	100	83.33	50	25	16.667	10	8.33	6.94	1	
1,728	864	576	172.8	144	86.4	43.2	28.8	17.28	14.4	12	1.728	1

159

Appendix B

Paper Quantities in the British Isles

each	quires	reams	bundles	bales
1				
25	1			
500	20	1		
1,000	40	2	1	
5,000	200	10	5	1

Appendix C

British Imperial Units

The tables on the following pages show the dimensions and the intra-unit proportions of standard weights and measures units in the British Isles since the passage of the Imperial Weights and Measures Act of 1824. There are no local variations recorded in any of the tables and the values given for each unit have remained constant.

Linear Measures

inches	Gunter's links	feet	yards	fathoms	rods	Gunter's chains	cable lengths	furlongs	miles	nautical miles	leagues
1											
7.92	1										
12	1.515	1									
36	4.545	3	1								
72	9.091	6	2	1							
198	25	16.5	5.5	2.75	1						
792	100	66	22	11	4	1					
7,200	909.091	600	200	100	36.36	9.091	1				
7,920	1,000	660	220	110	40	10	1.1	1			
63,360	8,000	5,280	1,760	880	320	80	8.8	8	1		
72,960	9,212.160	6,080	2,026.680	1,013.340	368.486	92.122	10.133	9.212	1.1515	1	
190,080	24,000	15,840	5,280	2,640	960	240	26.4	24	3	2.605	1

162

Area Measures

square inches	square feet	square yards	square rods	square Gunter's chains	roods	acres	square miles
1							
144	1						
1,296	9	1					
39,204	272.25	30.25	1				
627,264	4,356	484	16	1			
1,568,160	10,890	1,210	40	2.5	1		
6,272,640	43,560	4,840	160	10	4	1	
4,014,489,600	27,878,400	3,097,600	102,400	6,400	2,560	640	1

Liquid and Dry Capacity Measures

cubic inches	gills	pints	quarts	gallons	pecks	bushels	quarters	chalders
1								
8.669	1							
34.677	4	1						
69.355	8	2	1					
277.420	32	8	4	1				
554.840	64	16	8	2	1			
2,219.360	256	64	32	8	4	1		
17,754.880	2,048	512	256	64	32	8	1	
79,896.960	9,216	2,304	1,152	288	144	36	4.5	1

Apothecary Measures

cubic inches	minims	fluid drams	fluid ounces	pints	gallons
1					
0.0036122	1				
0.216734	60	1			
1.733875	480	8	1		
34.6775	9,600	160	20	1	
277.420	76,800	1,280	160	8	1

Culinary Measures

fluid ounces	fluid drams	tea-spoonsful	dessert-spoonsful	table-spoonsful	wine-glassesful	tea-cupsful	tum-blersful
1							
0.125	1						
0.125	1	1					
0.25	2	2	1				
0.50	4	4	2	1			
2.50	20	20	10	5	1		
5	40	40	20	10	2	1	
10	80	80	40	20	4	2	1

Avoirdupois Weights

troy grains	drams	ounces	pounds	stone	quarters	centals or short hundredweight	hundred-weight	short tons	long tons
1									
27.344	1								
437.5	16	1							
7,000	256	16	1						
98,000	3,584	224	14	1					
196,000	7,168	448	28	2	1				
700,000	25,600	1,600	100	7.142	3.571	1			
784,000	28,672	1,792	112	8	4	1.12	1		
14,000,000	512,000	32,000	2,000	142.856	71.428	20	17.857	1	
15,680,000	573,440	35,840	2,240	160	80	22.4	20	1.12	1

Apothecary Weights

troy grains	scruples	drams	ounces	pounds
1				
20	1			
60	3	1		
480	24	8	1	
5,760	288	96	12	1

Troy Weights

troy grains	pennyweights	ounces	pounds
1			
24	1		
480	20	1	
5,760	240	12	1

Appendix D

Pre-Metric Weights and Measures in Western and Eastern Europe

The following lists give the English imperial and metric equivalents for the principal standard weights and measures units of the major nations, provinces, and cities of western and eastern Europe from 1800 to the adoption of the metric system in each country. There are no local variations recorded in any of the lists. The five lists are arranged alphabetically by unit, and the imperial and metric equivalents, rounded off to two decimal places, or to three or four in cases of smaller units, remained constant throughout the period mentioned unless otherwise noted in parentheses. No French units are included since the metric system in France predates 1800. Whenever a unit name had more than one common transliteration given in a Western European or American source, I have used the most common one as the entry word and have placed the others in parentheses. The designation "Germany" indicates generalized use throughout all the German states.

List of Abbreviations

a	=	are or ares	mi	=	mile or miles
ac	=	acre or acres	mm	=	millimeter or millimeters
bu	=	bushel or bushels			
cm	=	centimeter or centimeters	oz	=	ounce or ounces
			pk	=	peck or pecks
ft	=	foot or feet	qt	=	quart or quarts
g	=	gram or grams	sq ft	=	square foot or feet
gal	=	gallon or gallons	sq km	=	square kilometer or kilometers
gr	=	grain or grains			
in	=	inch or inches	sq m	=	square meter or meters
km	=	kilometer or kilometers			
l	=	liter or liters	sq yd	=	square yard or yards
lb	=	pound or pounds	t	=	ton
m	=	meter or meters	yd	=	yard or yards

Appendix D

MEASURES OF LENGTH

Measure	Location	English Equivalent	Metric Equivalent
aln	Denmark	24.72 in	0.63 m
alen	Norway	23.38 in	0.59 m
archine (archinne, arshin)	Russia	28.00 in	0.71 m
arshien	Estonia	28.00 in	0.71 m
aune	Switzerland	47.24 in	1.20 m
braça	Portugal	7.22 ft	2.20 m
braccio	Ancona	25.33 in	0.64 m
	Genoa	20.95 in	0.53 m
	Modena	22.74 in	0.58 m
	Naples	27.51 in	0.70 m
	Parma	21.34 in	0.54 m
	Rome	30.73 in	0.78 m
	Tuscany	22.98 in	0.58 m
	Venice	27.38 in	0.69 m
brache	Switzerland	23.62 in	0.60 m
canna	Genoa	7.35 ft	2.24 m
	Malta	6.87 ft	2.09 m
	Naples	6.92 ft	2.11 m
	Rome	6.53 ft	1.99 m
	Sicily	6.35 ft	1.94 m
covado	Portugal	25.97 in	0.66 m
dedo	Portugal	0.72 in	0.018 m
	Spain	0.68 in	0.017 m
duim	Netherlands	0.39 in	0.010 m
duime (duim)	Russia	1.00 in	0.025 m
el	Netherlands	39.37 in	1.00 m
elle	Austria	30.68 in	0.78 m
	Baden	23.62 in	0.60 m
	Bavaria	32.22 in	0.82 m
	Bremen	22.77 in	0.58 m
	Brunswick	22.47 in	0.57 m
	Darmstadt	23.60 in	0.60 m
	Hamburg	22.54 in	0.57 m
	Hanover	22.99 in	0.58 m
	Kassel	22.43 in	0.57 m
	Lübeck	22.67 in	0.58 m
	Prague	23.20 in	0.59 m
	Prussia	26.26 in	0.67 m
	Saxony	22.30 in	0.57 m
	Stuttgart	24.18 in	0.61 m

Measures of Length (continued)

Measure	Location	English Equivalent	Metric Equivalent
endaseh (endazeh)	Rumania	26.08 in	0.66 m
estadal	Spain	11.13 ft	3.39 m
estadio	Portugal	0.16 mi	0.26 km
famn	Norway	70.14 in	1.78 m
favn	Denmark	74.15 in	1.88 m
fod	Denmark	12.36 in	0.31 m
fot	Norway	11.69 in	0.30 m
	Sweden	11.68 in	0.30 m
foute	Russia	12.00 in	0.30 m
fuss	Austria	12.45 in	0.32 m
	Baden	11.81 in	0.30 m
	Bavaria	11.49 in	0.29 m
	Bremen	11.39 in	0.29 m
	Darmstadt	9.84 in	0.25 m
(werkschuh)	Frankfurt	11.20 in	0.28 m
(feldfuss)	Frankfurt	14.01 in	0.36 m
	Hamburg	11.27 in	0.29 m
	Hanover	11.50 in	0.29 m
(waldfuss)	Kassel	11.32 in	0.29 m
	Lübeck	11.33 in	0.29 m
	Prague	11.88 in	0.30 m
(Rheinfuss)	Prussia	12.36 in	0.31 m
	Saxony	11.15 in	0.28 m
	Württemberg	11.26 in	0.29 m
grao	Portugal	0.18 in	0.005 m
grenzmiil	Norway	5.53 mi	8.92 km
jarda ionia	Ionian Islands	36.00 in	0.91 m
khalebi	Rumania	26.43 in	0.67 m
klafter	Austria	6.22 ft	1.90 m
	Baden	5.91 ft	1.80 m
	Bavaria	5.76 ft	1.76 m
	Bremen	5.69 ft	1.73 m
	Darmstadt	8.20 ft	2.50 m
	Württemberg	5.63 ft	1.72 m
kot	Rumania	24.86 in	0.63 m
lakat	Bulgaria	25.86 in	0.65 m
legoa	Portugal	3.84 mi	6.17 km
legua	Spain	4.21 mi and 3.46 mi	6.78 km 5.57 km
lieue	Switzerland	2.98 mi	4.80 km
ligne	Switzerland	0.12 in	0.003 m

Appendix D

Measures of Length (continued)

Measure	Location	English Equivalent	Metric Equivalent
linea	Spain	0.077 in	0.002 m
linha	Portugal	0.090 in	0.002 m
linia	Russia	0.083 in	0.002 m
linie	Austria	0.086 in	0.002 m
	Denmark	0.085 in	0.002 m
	Norway	0.081 in	0.002 m
	Sweden	0.12 in	0.003 m
lorkat	Estonia	21.00 in	0.53 m
meile	Austria	4.71 mi	7.59 km
	Baden	5.52 mi	8.89 km
	Bavaria	4.61 mi	7.43 km
	Bremen	3.90 mi	6.28 km
	Brunswick	4.60 mi	7.42 km
	Hamburg	4.68 mi	7.53 km
	Hanover	4.61 mi	7.43 km
	Lithuania	5.00 mi	8.04 km
	Prussia	4.68 mi	7.53 km
(meile post)	Saxony	4.22 mi	6.80 km
(meile polizei)	Saxony	5.63 mi	9.06 km
meripeninkulma	Finland	1.15 mi	1.85 km
miglio	Lombardy	0.62 mi	1.00 km
	Naples	1.15 mi	1.85 km
	Sardinia	1.38 mi	2.23 km
	Tuscany	1.03 mi	1.65 km
miil (mil)	Denmark	4.68 mi	7.53 km
miil	Norway	6.64 mi	10.69 km
	Sweden	6.21 mi	10.00 km
milha	Portugal	1.28 mi	2.06 km
palm	Netherlands	3.94 in	0.10 m
palma	Rumania	10.89 in	0.28 m
palmo	Genoa	8.98 in	0.23 m
	Naples	10.38 in	0.26 m
	Portugal	8.66 in	0.22 m
	Rome	8.80 in	0.22 m
	Sicily	9.53 in	0.25 m
	Spain	8.35 in	0.21 m
	Tuscany	11.49 in	0.29 m
peninkulma	Finland	6.21 mi	10.00 km
perche	Switzerland	9.84 ft	3.00 m
pic (pik)	Crete	25.09 in	0.64 m
(pike)	Cyprus	26.45 in	0.67 m
	Greece	27.00 in	0.69 m

Measures of Length (continued)

Measure	Location	English Equivalent	Metric Equivalent
pie	Spain	11.13 in	0.28 m
pied	Switzerland	11.81 in	0.30 m
piede	Ancona	15.38 in	0.39 m
(liprando)	Genoa	20.23 in	0.51 m
(manual)	Genoa	13.48 in	0.34 m
	Milan	15.61 in	0.40 m
	Modena	20.59 in	0.52 m
	Parma	22.43 in	0.57 m
	Rome	11.59 in	0.29 m
	Venice	13.69 in	0.35 m
pollegada	Portugal	1.08 in	0.03 m
ponto	Portugal	0.007 in	0.0002 m
pouce	Switzerland	1.18 in	0.03 m
pulgada	Spain	0.91 in	0.023 m
punkt	Austria	0.007 in	0.0002 m
punto	Spain	0.006 in	0.0002 m
ref	Sweden	97.32 ft	29.66 m
rode	Denmark	10.30 ft	3.14 m
	Norway	9.74 ft	2.97 m
roede	Netherlands	32.81 ft	10.00 m
ruthe	Austria	12.45 ft	3.79 m
	Baden	9.84 ft	3.00 m
	Bavaria	9.57 ft	2.92 m
	Bremen	15.18 ft	4.63 m
	Brunswick	14.98 ft	4.57 m
	Frankfurt	11.67 ft	3.56 m
	Hanover	15.33 ft	4.67 m
	Kassel	13.09 ft	3.99 m
	Prussia	12.36 ft	3.77 m
	Saxony	14.86 ft	4.53 m
	Württemberg	9.38 ft	2.86 m
sachine (sagene, sajene, sashen)	Russia	7.00 ft	2.13 m
schuh	Brunswick	11.23 in	0.28 m
sesma	Spain	5.56 in	0.14 m
stab	Hungary	62.23 in	1.58 m
stadio	Ionian Islands	66.00 ft	20.12 m
stang	Sweden	9.73 ft	2.97 m
streep	Netherlands	0.04 in	1.00 mm
tchetverk	Russia	7.00 in	0.18 m
toise	Switzerland	11.81 ft	3.60 m
tomme	Denmark	1.03 in	0.03 m

Measures of Length (continued)

Measure	Location	English Equivalent	Metric Equivalent
trait	Switzerland	0.012 in	0.0003 m
tum	Norway	0.97 in	0.025 m
	Sweden	1.17 in	0.03 m
vadem	Netherlands	2.00 yd	1.83 m
vara	Portugal	43.28 in	1.10 m
		43.74 in (1934)	1.11 m
		43.31 in (1956)	1.10 m
	Spain	33.38 in	0.85 m
		32.91 in (1934)	0.84 m
		33.14 in (1956)	0.84 m
vershock	Estonia	1.75 in	0.04 m
vershok	Russia	1.75 in	0.04 m
verst (versta, werst)	Russia	0.66 mi	1.07 km
voet	Netherlands	0.93 ft	28.30 cm
wegstunde	Baden	2.76 mi	4.44 km
zoll	Austria	1.04 in	0.026 m
	Baden	1.18 in	0.030 m
	Bavaria	0.96 in	0.024 m
	Bremen	0.95 in	0.024 m
	Brunswick	0.94 in	0.024 m
	Frankfurt	0.93 in	0.024 m
	Hamburg	0.94 in	0.024 m
	Hanover	0.96 in	0.024 m
	Kassel	0.94 in	0.024 m
	Lübeck	0.94 in	0.024 m
	Prussia	1.03 in	0.026 m
	Saxony	0.93 in	0.024 m
	Württemberg	1.13 in	0.029 m

MEASURES OF AREA

Measure	Location	English Equivalent	Metric Equivalent
acker	Kassel	0.59 ac	23.86 a
arpent	Switzerland	3.56 ac	144.00 a
biolca	Modena	0.70 ac	28.36 a
	Parma	0.75 ac	30.46 a
bunder	Netherlands	2.47 ac	100.00 a
cadastral yoke	Hungary	1.42 ac	57.55 a
campo	Venetian Lombardy	0.69 ac	27.84 a

Measures of Area (continued)

Measure	Location	English Equivalent	Metric Equivalent
celemin	Spain	641.80 sq yd	536.63 sq m
deciatina (dessiatina)	Russia	2.70 ac	119.25 a
dessiatine	Estonia	2.70 ac	119.25 a
donum	Cyprus	0.33 ac	13.44 a
drohn	Hanover	0.48 ac	19.66 a
fanegada (fanega)	Spain	1.59 ac	64.22 a
fjerding	Norway	0.15 ac	6.13 a
geira	Portugal	1.43 ac	57.81 a
giornata	Sardinia	0.94 ac	38.01 a
jitro	Czechoslovakia	1.58 ac	63.82 a
joch	Austria	1.42 ac	57.55 a
jutro	Yugoslavia	1.42 ac	57.55 a
kappland	Norway	0.04 ac	1.53 a
kastastralis hold	Hungary	1.55 ac	62.73 a
moggio	Naples	0.86 ac	34.73 a
morga	Poland	1.35 ac	54.63 a
morgen	Baden	0.89 ac	36.03 a
	Bavaria	0.84 ac	34.08 a
	Bremen	0.63 ac	25.69 a
	Brunswick	0.62 ac	25.14 a
	Darmstadt	0.62 ac	25.00 a
	Frankfurt	0.50 ac	20.25 a
	Hamburg	2.38 ac	96.51 a
	Hanover	0.65 ac	26.19 a
	Prussia	0.63 ac	25.55 a
	Saxony	1.52 ac	61.62 a
	Württemberg	0.78 ac	31.54 a
negyszögöl	Hungary	4.68 sq yd	3.91 sq m
ref	Sweden	0.22 ac	8.80 a
rubbio	Rome	4.57 ac	184.75 a
saccato	Tuscany	1.39 ac	56.20 a
spannland	Norway	0.61 ac	24.54 a
square arshin	Russia	5.44 sq ft	0.51 sq m
square aune	Switzerland	15.50 sq ft	1.44 sq m
square fod	Denmark	1.06 sq ft	0.098 sq m
square fot	Norway	0.95 sq ft	0.088 sq m
	Sweden	0.95 sq ft	0.088 sq m
square foute	Russia	1.00 sq ft	0.093 sq m
square fuss	Austria	1.08 sq ft	0.099 sq m
	Baden	0.97 sq ft	0.090 sq m

Appendix D

Measures of Area (continued)

Measure	Location	English Equivalent	Metric Equivalent
	Bavaria	0.92 sq ft	0.085 sq m
	Bremen	0.90 sq ft	0.084 sq m
	Darmstadt	0.67 sq ft	0.062 sq m
(werkschuh)	Frankfurt	0.87 sq ft	0.081 sq m
(feldfuss)	Frankfurt	1.36 sq ft	0.127 sq m
	Hamburg	0.88 sq ft	0.082 sq m
	Hanover	0.92 sq ft	0.085 sq m
(waldfuss)	Kassel	0.89 sq ft	0.083 sq m
	Lübeck	0.89 sq ft	0.083 sq m
(Rheinfuss)	Prussia	1.06 sq ft	0.098 sq m
	Saxony	0.86 sq ft	0.080 sq m
	Württemberg	0.88 sq ft	0.082 sq m
square klafter	Austria	38.72 sq ft	3.60 sq m
	Baden	34.88 sq ft	3.24 sq m
	Bavaria	33.23 sq ft	3.07 sq m
	Württemberg	31.70 sq ft	2.94 sq m
square palmo	Portugal	0.52 sq ft	0.066 sq m
	Spain	0.48 sq ft	0.045 sq m
square pie	Spain	0.86 sq ft	0.080 sq m
square pied	Switzerland	0.97 sq ft	0.090 sq m
square piede	Ancona	1.64 sq ft	0.153 sq m
(liprando)	Genoa	2.84 sq ft	0.264 sq m
(manual)	Genoa	1.26 sq ft	0.117 sq m
	Milan	1.69 sq ft	0.157 sq m
	Modena	2.94 sq ft	0.274 sq m
	Parma	3.49 sq ft	0.325 sq m
	Rome	0.93 sq ft	0.087 sq m
	Venice	1.30 sq ft	0.121 sq m
square rode	Denmark	106.05 sq ft	9.85 sq m
	Norway	94.91 sq ft	8.82 sq m
square ruthe	Austria	154.89 sq ft	14.39 sq m
	Baden	96.88 sq ft	9.00 sq m
	Bavaria	91.68 sq ft	8.52 sq m
	Bremen	230.46 sq ft	21.41 sq m
	Brunswick	224.40 sq ft	20.85 sq m
	Frankfurt	136.23 sq ft	12.55 sq m
	Hanover	234.95 sq ft	21.83 sq m
	Kassel	171.26 sq ft	15.91 sq m
	Prussia	152.70 sq ft	14.18 sq m
	Saxony	220.97 sq ft	20.52 sq m
	Württemberg	88.05 sq ft	8.18 sq m

Measures of Area (continued)

Measure	Location	English Equivalent	Metric Equivalent
square sachine (sashen)	Russia	49.00 sq ft	4.55 sq m
square schuh	Brunswick	0.88 sq ft	0.081 sq m
square vara	Portugal	13.01 sq ft	1.21 sq m
	Spain	7.74 sq ft	0.72 sq m
square verst	Russia	0.44 sq mi	1.14 sq km
square zoll	Austria	1.08 sq in	0.0007 sq m
	Baden	1.39 sq in	0.0009 sq m
	Bavaria	0.92 sq in	0.0006 sq m
	Bremen	0.90 sq in	0.0006 sq m
	Brunswick	0.88 sq in	0.0006 sq m
	Frankfurt	0.87 sq in	0.0006 sq m
	Hamburg	0.88 sq in	0.0006 sq m
	Hanover	0.92 sq in	0.0006 sq m
	Kassel	0.89 sq in	0.0006 sq m
	Lübeck	0.89 sq in	0.0006 sq m
	Prussia	1.06 sq in	0.0007 sq m
	Saxony	0.86 sq in	0.0005 sq m
	Württemberg	1.27 sq in	0.0008 sq m
stremma	Greece	0.25 ac	10.00 a
strich	Bohemia	0.71 ac	28.77 a
tavola	Modena	0.010 ac	0.394 a
	Parma	0.010 ac	0.394 a
töndeland	Denmark	1.36 ac	55.17 a
	Norway	1.21 ac	49.09 a
tornatura	Venetian Lombardy	0.025 ac	1.000 a
vierkante roede	Netherlands	0.025 ac	1.000 a
viertel	Baden	0.22 ac	9.00 a
	Darmstadt	0.15 ac	6.25 a
vorling	Hanover	0.33 ac	13.10 a
yugada	Spain	79.35 ac	3,210.94 a

MEASURES OF CAPACITY FOR LIQUIDS

Measure	Location	English Equivalent	Metric Equivalent
almud	Lisbon	3.64 gal	16.54 l
almude	Oporto	5.52 gal	25.08 l
alqueire	Lisbon	1.82 gal	8.27 l
	Oporto	2.76 gal	12.54 l

Appendix D

Measures of Capacity for Liquids (continued)

Measure	Location	English Equivalent	Metric Equivalent
amola	Genoa	0.16 gal	0.72 l
anker	Bremen	7.97 gal	36.24 l
	Denmark	8.29 gal	37.67 l
	Hamburg	7.94 gal	36.20 l
	Hanover	8.63 gal	38.94 l
	Lübeck	8.24 gal	37.46 l
	Prussia	7.56 gal	34.35 l
	Saxony	14.31 gal	65.04 l
arroba	Spain (wine)	3.55 gal	16.14 l
	(oil)	2.76 gal	12.56 l
azumbre	Spain	0.44 gal	2.02 l
barile	Ionian Islands	16.00 gal	72.71 l
	Malta (wine)	9.15 gal	41.63 l
	Modena (wine)	9.17 gal	41.66 l
	Naples (wine)	9.64 gal	43.81 l
	Rome (wine)	12.84 gal	58.34 l
	Sardinia	14.23 gal	64.67 l
	Tuscany (wine)	10.03 gal	45.59 l
boccale	Modena	0.23 gal	1.04 l
	Rome (wine)	0.40 gal	1.82 l
	Sardinia	0.17 gal	0.78 l
	Tuscany	0.25 gal	1.14 l
boot	Netherlands (sherry)	116.60 gal	530.00 l
bota	Portugal	94.64 gal	430.13 l
botta	Naples (wine)	107.34 gal	487.84 l
	Rome (wine)	205.39 gal	933.42 l
caffiso	Malta (oil)	4.58 gal	20.81 l
canada	Portugal	0.30 gal	1.48 l
caraffa	Naples	0.64 qt	0.73 l
carro	Naples	214.69 gal	975.69 l
	Sardinia	123.96 gal	563.34 l
cass	Cyprus	1.41 gal	4.73 l
copa	Spain	0.11 qt	0.13 l
coppo	Venetian Lombardy	0.88 qt	1.00 l
cuartillo	Spain	0.44 qt	0.51 l
dicotoli	Ionian Islands	0.50 qt	0.57 l
driling	Austria	298.90 gal	1,358.40 l
eimer	Austria	12.45 gal	56.60 l
	Bavaria (wine)	14.11 gal	64.14 l
	Hamburg	6.35 gal	28.88 l
	Hanover	13.81 gal	62.74 l
	Lübeck	6.59 gal	29.96 l

Pre-Metric European Weights and Measures

Measures of Capacity for Liquids (continued)

Measure	Location	English Equivalent	Metric Equivalent
	Prague	13.44 gal	61.09 l
	Prussia	15.11 gal	68.69 l
	Saxony	16.69 gal	75.85 l
	Württemberg	64.67 gal	293.93 l
fass	Brunswick (beer)	88.88 gal	403.91 l
	Hanover	89.74 gal	407.84 l
	Hungary	43.77 gal	198.90 l
	Prague	53.77 gal	244.38 l
	Saxony	83.44 gal	379.19 l
fiasco	Modena	1.83 qt	2.08 l
	Tuscany	2.01 qt	2.28 l
fuder	Austria	398.53 gal	1,811.19 l
	Baden	336.06 gal	1,500.00 l
	Bremen	191.39 gal	869.78 l
	Brunswick	197.50 gal	897.57 l
	Darmstadt	211.24 gal	960.00 l
	Frankfurt	189.36 gal	860.06 l
	Hamburg	190.64 gal	866.39 l
	Hanover	207.09 gal	941.17 l
	Kassel	209.50 gal	952.09 l
	Lübeck	197.77 gal	898.80 l
	Prussia	181.40 gal	824.39 l
	Saxony	200.28 gal	910.22 l
	Württemberg	209.50 gal	952.09 l
glas	Baden	0.13 qt	0.15 l
halbstück	Netherlands (Rhine wine)	132.00 gal	600.00 l
jungfru	Norway	0.072 qt	0.082 l
kan	Netherlands	0.88 qt	1.00 l
kande	Denmark	1.70 qt	1.93 l
kanna	Norway	2.30 qt	2.62 l
	Sweden	2.30 qt	2.62 l
kanne	Austria	0.62 qt	0.71 l
	Hamburg	1.59 qt	1.81 l
	Hanover	1.73 qt	1.96 l
	Lübeck	1.65 qt	1.87 l
	Saxony	1.06 qt	1.20 l
maatje	Netherlands	0.088 qt	0.100 l
mass	Austria	1.24 qt	1.41 l
	Baden	1.32 qt	1.50 l
	Bavaria (beer)	0.94 qt	1.07 l
	Darmstadt	1.76 qt	2.00 l
(eichmass)	Frankfurt	1.58 qt	1.79 l

Appendix D

Measures of Capacity for Liquids (continued)

Measure	Location	English Equivalent	Metric Equivalent
(neumass)	Frankfurt	1.40 qt	1.59 l
	Kassel (wine)	1.75 qt	1.98 l
	Württemberg	1.62 qt	1.84 l
mezzaruola	Sardinia	28.46 gal	129.34 l
mezzetta	Tuscany	0.50 qt	0.57 l
mina	Venetian Lombardy	2.20 gal	10.00 l
mingel	Bremen	0.18 qt	0.20 l
mistate	Crete	2.45 gal	11.16 l
misurella	Naples (oil)	0.093 qt	0.105 l
muid	Switzerland	33.01 gal	150.00 l
nössel	Brunswick	0.41 qt	0.47 l
	Hamburg	0.40 qt	0.45 l
	Hanover	0.43 qt	0.49 l
	Prussia	0.50 qt	0.57 l
	Saxony	0.55 qt	0.60 l
ohm	Baden	33.01 gal	150.00 l
	Bremen	31.90 gal	144.96 l
	Brunswick	32.92 gal	149.60 l
	Darmstadt	35.21 gal	160.00 l
	Frankfurt	31.56 gal	143.43 l
	Hamburg	31.77 gal	144.91 l
	Hanover	34.52 gal	155.76 l
	Kassel (beer)	38.42 gal	174.63 l
	Lübeck	32.13 gal	149.80 l
	Prussia	30.23 gal	137.38 l
	Saxony	33.38 gal	151.72 l
oka	Rumania	1.25 qt	1.42 l
oke	Crete	1.15 qt	1.31 l
okshoofd	Netherlands	51.20 gal	232.80 l
ort	Lübeck	0.21 qt	0.23 l
oxhoft	Bremen	49.85 gal	227.45 l
	Brunswick	49.37 gal	224.39 l
	Hamburg	47.66 gal	216.60 l
	Hanover	51.77 gal	235.29 l
	Lübeck	49.44 gal	224.70 l
	Saxony	50.07 gal	227.55 l
panilla	Spain	0.11 qt	0.13 l
pfiff	Austria	0.16 qt	0.18 l
pinta	Sardinia	1.38 qt	1.57 l
	Venetian Lombardy	0.88 qt	1.00 l
pipa	Madeira (wine)	91.62 gal	416.38 l
	Portugal	47.31 gal	215.02 l

Pre-Metric European Weights and Measures

Measures of Capacity for Liquids (continued)

Measure	Location	English Equivalent	Metric Equivalent
planke	Lübeck	0.41 qt	0.47 l
poegel	Denmark	0.21 qt	0.24 l
pot	Denmark	0.85 qt	0.97 l
	Switzerland	1.32 qt	1.50 l
quarter	Norway	0.29 qt	0.33 l
quartier	Bremen	0.71 qt	0.81 l
	Brunswick	0.82 qt	0.93 l
	Hamburg	0.79 qt	0.90 l
	Hanover	0.86 qt	0.98 l
	Lübeck	0.82 qt	0.93 l
	Prussia	1.01 qt	1.14 l
	Saxony	0.43 qt	0.49 l
quartilho	Portugal	0.30 qt	0.34 l
quartillo	Spain	0.44 qt	0.49 l
quartino	Sardinia	0.34 qt	0.39 l
quarto	Naples (oil)	0.54 qt	0.62 l
quartuccio	Tuscany	0.25 qt	0.28 l
rubbio	Piedmont	2.07 gal	9.39 l
salma	Sicily	18.32 gal	83.28 l
schoppen	Darmstadt	0.44 qt	0.50 l
	Frankfurt	0.39 qt	0.45 l
	Kassel	0.44 qt	0.50 l
	Württemberg	0.40 qt	0.46 l
seidel	Austria	0.31 qt	0.35 l
	Prague	0.42 qt	0.48 l
setier	Switzerland	8.25 gal	37.50 l
soma	Tuscany (oil)	16.05 gal	72.95 l
	Venetian Lombardy	22.00 gal	100.00 l
staio	Naples (oil)	2.18 gal	9.90 l
stoop	Netherlands (Pils beer)	1.32 gal	6.00 l
stop	Norway	1.15 qt	1.31 l
stübchen	Bremen	2.84 qt	3.23 l
	Brunswick	3.29 qt	3.74 l
	Hamburg	3.18 qt	3.61 l
	Hanover	3.45 qt	3.93 l
	Lübeck	3.30 qt	3.75 l
stütz	Baden	3.30 gal	15.00 l
tönde	Denmark (beer)	28.91 gal	131.38 l
	(tar)	25.51 gal	115.92 l
tonelada	Portugal	94.62 gal	430.00 l
tonne	Brunswick	22.22 gal	100.98 l

Measures of Capacity for Liquids (continued)

Measure	Location	English Equivalent	Metric Equivalent
	Hamburg	38.13 gal	173.28 l
	Hanover	34.37 gal	156.16 l
tunna	Norway	27.63 gal	125.56 l
vat	Netherlands	22.00 gal	100.00 l
vedro	Russia	2.71 gal	12.30 l
viadra	Rumania	3.12 gal	14.17 l
viertel	Austria	3.11 gal	14.15 l
	Bremen	1.59 gal	7.21 l
	Darmstadt	1.76 gal	8.00 l
	Frankfurt	1.58 gal	7.17 l
	Hamburg	1.59 gal	7.21 l
	Hanover	1.73 gal	7.84 l
	Kassel	1.75 gal	7.94 l
	Lübeck	1.65 gal	7.49 l
vingerhoed	Netherlands	0.009 qt	0.010 l

MEASURES OF CAPACITY FOR DRY PRODUCTS

Measure	Location	English Equivalent	Metric Equivalent
achtel	Austria	0.85 pk	7.69 l
	Württemberg	0.30 pk	3.65 l
almud (almude)	Spain	0.52 pk	4.66 l
alquiere	Portugal	1.49 pk	13.52 l
becher	Austria	0.053 pk	0.479 l
	Baden	0.016 pk	0.150 l
	Brunswick	0.21 pk	1.91 l
cahiz	Spain	18.60 bu	676.20 l
carga	Crete	4.19 bu	152.33 l
carro	Naples	54.61 bu	1,984.70 l
chilo	Ionian Islands	1.00 bu	36.35 l
corba	Bologna	2.15 bu	78.63 l
dreissiger	Bavaria	0.13 pk	1.16 l
drittel	Hanover	1.14 pk	10.39 l
drömt	Lübeck	11.75 bu	426.93 l
ecklein	Württemberg	0.006 pk	0.069 l
emine	Switzerland	0.16 pk	1.50 l
fanega	Spain	1.55 bu	56.34 l
fanga	Lisbon	1.49 bu	54.08 l
	Oporto	1.88 bu	62.27 l
fass	Hamburg	1.45 bu	52.65 l

Pre-Metric European Weights and Measures

Measures of Capacity for Dry Products (continued)

Measure	Location	English Equivalent	Metric Equivalent
	Lübeck	0.98 pk	8.89 l
fjerding	Norway	2.01 pk	18.31 l
fjerdingkar	Denmark	0.96 pk	8.69 l
garnetz	Russia	0.38 pk	3.46 l
gombetta	Genoa	1.38 pk	12.57 l
grosses massel	Austria	0.21 pk	1.92 l
halvotting	Denmark	0.24 pk	2.17 l
himt	Brunswick	3.36 pk	30.54 l
	Hamburg	2.90 pk	26.32 l
	Hanover	3.43 pk	31.16 l
	Kassel	4.42 pk	40.18 l
kanna	Norway	0.29 pk	2.61 l
	Sweden	2.88 pk	26.17 l
kappa (kappe)	Finland	4.40 qt	5.00 l
kleines massel	Austria	0.11 pk	1.00 l
kop	Netherlands	0.11 pk	1.00 l
kümpf	Darmstadt	0.88 pk	8.00 l
last	Bremen	81.51 bu	2,962.57 l
	Hamburg	86.91 bu	3,158.68 l
	Hanover	82.30 bu	2,991.17 l
	Lübeck	93.97 bu	3,415.46 l
	Netherlands	82.54 bu	3,000.00 l
	Prussia	108.86 bu	3,956.72 l
maatje	Netherlands	0.011 pk	0.100 l
mässchen	Darmstadt	0.055 pk	0.500 l
	Frankfurt	0.049 pk	0.448 l
	Kassel	0.28 pk	2.53 l
	Prussia	0.094 pk	0.859 l
	Saxony	0.18 pk	1.62 l
mässlein	Baden	0.16 pk	1.50 l
	Bavaria	0.25 pk	2.32 l
	Württemberg	0.15 pk	1.38 l
malter	Baden	4.13 bu	150.00 l
	Darmstadt	3.52 bu	128.00 l
	Frankfurt	3.16 bu	114.74 l
	Hanover	5.14 bu	186.95 l
	Prussia	18.14 bu	659.49 l
	Saxony	34.30 bu	1,246.70 l
maquia	Portugal	0.093 pk	0.845 l
massel	Bavaria	0.51 pk	4.64 l
	Prague	0.64 pk	5.85 l
medinno	Cyprus	2.07 bu	75.09 l

Appendix D

Measures of Capacity for Dry Products (continued)

Measure	Location	English Equivalent	Metric Equivalent
medio	Spain	0.25 pk	2.35 l
metze	Austria	1.69 bu	61.49 l
	Bavaria	4.08 pk	37.06 l
	Frankfurt	1.58 pk	14.34 l
	Hungary	1.72 bu	62.49 l
	Kassel	1.11 pk	10.13 l
	Prussia	0.38 pk	3.43 l
	Saxony	0.71 pk	6.49 l
mina	Parma	0.65 bu	23.50 l
	Sardinia	3.32 bu	120.71 l
	Turin	0.63 bu	23.00 l
	Tuscany	1.61 pk	14.61 l
	Venetian Lombardy	1.10 pk	10.00 l
misura	Naples	0.25 pk	2.30 l
moggio	Tuscany	16.08 bu	584.58 l
moio	Portugal	22.32 bu	811.14 l
mudde	Netherlands	2.75 bu	100.00 l
müller-massel	Austria	0.42 pk	3.84 l
muth	Austria	50.75 bu	1,844.73 l
ochavillo	Spain	0.008 pk	0.074 l
ort	Norway	0.009 pk	0.082 l
osmina	Russia	2.89 bu	104.95 l
ottingkar	Denmark	0.96 pk	8.69 l
outava	Portugal	0.19 pk	1.69 l
pajak	Russia	1.44 bu	52.47 l
quarta	Rome	2.02 bu	73.60 l
quarter	Norway	0.038 pk	0.343 l
quarterolo	Parma	0.32 pk	2.94 l
quarteron	Switzerland	1.65 pk	15.00 l
quarterone	Bologna	1.08 pk	9.83 l
quarticino	Bologna	0.27 pk	2.46 l
quartillo	Spain	0.13 pk	1.17 l
quarto	Genoa	1.66 pk	15.09 l
	Portugal	0.37 pk	3.38 l
quartuccio	Rome	0.37 pk	3.35 l
racion	Spain	0.032 pk	0.293 l
rubbio	Ancona	7.73 bu	280.97 l
	Rome	8.10 bu	294.40 l
sacco	Modena	3.87 bu	140.80 l
	Turin	3.16 bu	114.99 l
	Tuscany	2.01 bu	73.07 l
salma	Malta	7.97 bu	289.78 l

Pre-Metric European Weights and Measures

Measures of Capacity for Dry Products (continued)

Measure	Location	English Equivalent	Metric Equivalent
(generale)	Sicily	7.56 bu	274.84 l
(grosso)	Sicily	9.69 bu	353.26 l
scheepston	Netherlands	27.51 bu	1,000.00 l
scheffel	Bavaria	6.12 bu	222.34 l
	Bremen	2.04 bu	74.06 l
	Brunswick	8.40 bu	305.40 l
	Hamburg	2.90 bu	105.30 l
	Kassel	2.23 bu	81.25 l
	Lübeck (wheat)	0.98 bu	35.58 l
	(oats)	1.09 bu	39.51 l
	Prussia	1.51 bu	54.96 l
	Saxony	2.86 bu	103.89 l
	Württemberg	4.88 bu	177.21 l
schepel	Netherlands	1.10 pk	10.00 l
schrott	Frankfurt	0.012 pk	0.112 l
scorzo	Rome	1.47 pk	13.38 l
sechter	Frankfurt	0.79 pk	7.17 l
seidel	Prague	0.054 pk	0.488 l
sester	Baden	1.65 pk	15.00 l
simmer	Darmstadt	3.52 pk	32.00 l
	Frankfurt	3.16 pk	28.68 l
simri	Württemberg	2.44 pk	22.15 l
skoeppe	Denmark	1.91 pk	17.39 l
soma	Venetian Lombardy	2.75 bu	100.00 l
span	Norway	2.01 bu	73.24 l
spint	Bremen	0.51 pk	4.63 l
	Hamburg	0.72 pk	6.58 l
staio	Bologna	1.08 bu	39.32 l
	Modena	1.94 bu	70.39 l
	Parma	1.29 bu	46.99 l
	Turin	1.05 bu	38.33 l
	Tuscany	0.67 bu	24.37 l
starello	Rome	2.02 pk	18.40 l
stop	Norway	0.15 pk	1.37 l
strich	Prague	2.57 bu	93.59 l
tchetverik (chetverik)	Russia	2.89 pk	26.24 l
tchetverka	Russia	0.72 pk	6.56 l
tchetviert (chetvert)	Russia	5.77 bu	209.73 l
tomolo	Naples	1.52 bu	55.13 l
tönde	Denmark	3.83 bu	139.10 l

Measures of Capacity for Dry Products (continued)

Measure	Location	English Equivalent	Metric Equivalent
tonne	Lübeck	3.91 bu	142.31 l
tunna	Norway	4.03 bu	146.48 l
vierfass	Hanover	0.86 pk	7.79 l
viertel	Austria	1.69 pk	15.37 l
	Bavaria	2.04 pk	18.53 l
	Bremen	2.04 pk	18.53 l
	Prague	2.57 pk	23.40 l
	Prussia	1.51 pk	13.74 l
	Saxony	2.86 pk	25.97 l
	Württemberg	0.61 pk	5.54 l
viertlein	Württemberg	0.002 pk	0.017 l
wispel	Brunswick	33.61 bu	1,221.59 l
	Hamburg	28.97 bu	1,052.89 l
	Hanover	41.15 bu	1,495.59 l
	Prussia	27.22 bu	989.82 l
	Saxony	68.48 bu	2,493.45 l
zuber	Baden	41.27 bu	1,500.00 l

WEIGHTS

Measure	Location	English Equivalent	Metric Equivalent
accino	Naples	0.69 gr	0.045 g
adarme	Spain	27.78 gr	1.80 g
arratel	Portugal	1.01 lb	460.00 g
arroba	Portugal	32.38 lb	14,687.40 g
	Spain	25.40 lb	11,522.60 g
as	Baden	0.77 gr	0.050 g
	Norway	0.74 gr	0.048 g
berkovetz (berkowitz)	Russia	360.68 lb	163,603.00 g
cantaro	Crete	116.57 lb	52,886.00 g
	Cyprus	524.19 lb	237,774.00 g
	Ionian Islands	118.80 lb	53,888.00 g
	Malta	175.00 lb	79,381.00 g
	Rumania	123.99 lb	56,243.00 g
carato	Bologna	2.91 gr	0.19 g
centinajo	Ionian Islands	100.00 lb	45,360.00 g
dekas	Baden	7.77 gr	0.50 g
denaro	Genoa	17.02 gr	1.10 g
	Parma	17.49 gr	1.13 g

Weights (continued)

Measure	Location	English Equivalent	Metric Equivalent
	Rome	18.17 gr	1.18 g
	Turin	19.77 gr	1.28 g
	Tuscany	18.19 gr	1.18 g
	Venetian Lombardy	15.43 gr	1.00 g
doli	Russia	0.68 gr	0.044 g
drachma	Greece	15.43 gr	1.00 g
	Russia	57.60 gr	3.73 g
drachme	Austria	67.70 gr	4.39 g
	Denmark	57.53 gr	3.73 g
	Germany	57.53 gr	3.73 g
dramma	Naples	41.25 gr	2.67 g
	Tuscany	54.59 gr	3.54 g
ferlino	Bologna	29.11 gr	1.89 g
	Modena	25.68 gr	1.66 g
frachtspfund	Bremen	329.72 lb	149,560.00 g
funte	Russia	0.903 lb	409.51 g
gran	Austria (apothecary)	1.13 gr	0.073 g
	Denmark (apothecary)	0.96 gr	0.062 g
	Germany (apothecary)	0.96 gr	0.062 g
	Norway	0.95 gr	0.061 g
	Russia	0.96 gr	0.062 g
	Saxony	0.94 gr	0.061 g
grano	Bologna	0.72 gr	0.047 g
	Genoa	0.71 gr	0.046 g
	Parma	0.73 gr	0.047 g
	Rome	0.76 gr	0.049 g
	Spain	0.77 gr	0.050 g
	Turin	0.82 gr	0.053 g
	Tuscany	0.76 gr	0.049 g
	Venetian Lombardy	1.54 gr	0.100 g
granotino	Turin	0.030 gr	0.002 g
grao	Portugal	0.77 gr	0.050 g
grosso	Venetian Lombardy	154.32 gr	10.00 g
heller	Brunswick	7.04 gr	0.46 g
	Frankfurt	7.62 gr	0.49 g
karaatti	Finland	3.09 gr	0.20 g
karch	Austria	493.85 lb	224,009.00 g
korrel	Netherlands	1.54 gr	0.100 g
korzec	Poland	216.00 lb	97,975.87 g

Appendix D

Weights (continued)

Measure	Location	English Equivalent	Metric Equivalent
lana	Russia	1.204 oz	34.12 g
last	Austria	4,938.46 lb	2,250,085.00 g
	Sweden	9,337.90 lb	4,245,671.00 g
libbra	Ancona	0.73 lb	350.53 g
	Bologna	0.80 lb	362.15 g
(sottile)	Genoa	0.70 lb	317.61 g
(grosso)	Genoa	0.77 lb	348.85 g
(sottile)	Ionian Islands	0.82 lb	373.27 g
(grosso)	Ionian Islands	1.00 lb	453.60 g
	Malta	1.75 lb	794.00 g
	Modena	0.70 lb	319.52 g
	Naples	0.71 lb	320.79 g
	Parma	0.72 lb	326.46 g
	Rome	0.75 lb	339.16 g
	Sicily	0.71 lb	320.79 g
	Turin	0.81 lb	368.91 g
	Tuscany	0.75 lb	339.16 g
libra	Spain	1.01 lb	460.00 g
mazsa	Hungary	220.46 lb	99,998.89 g
oncia	Modena	0.939 oz	26.63 g
	Naples	0.943 oz	26.73 g
	Parma	0.960 oz	27.20 g
	Rome	0.997 oz	28.06 g
	Sicily	0.943 oz	26.73 g
	Turin	1.084 oz	30.74 g
	Tuscany	0.998 oz	28.30 g
	Venetian Lombardy	3.527 oz	100.00 g
ons	Netherlands	3.527 oz	100.00 g
onza	Spain	1.106 oz	28.81 g
ort	Bremen	15.03 gr	0.97 g
	Denmark (ordinary)	77.16 gr	5.00 g
	(gold and silver)	14.08 gr	0.91 g
	Sweden	65.35 gr	4.23 g
ottavo	Bologna	58.22 gr	3.77 g
	Turin	59.30 gr	3.84 g
outava	Portugal	55.34 gr	3.59 g
packen	Russia	1,082.03 lb	490,810.00 g
pfund	Austria	1.235 lb	560.00 g
	Baden	1.102 lb	500.00 g
	Bavaria	1.102 lb	500.00 g
	Bremen	1.099 lb	498.54 g
	Brunswick	1.031 lb	467.48 g

Weights (continued)

Measure	Location	English Equivalent	Metric Equivalent
	Darmstadt	1.102 lb	500.00 g
	Frankfurt	1.114 lb	505.34 g
	Hamburg	1.068 lb	484.40 g
	Hanover	1.079 lb	489.64 g
	Kassel	1.067 lb	484.22 g
	Lübeck	1.069 lb	484.77 g
	Prussia	1.102 lb	500.00 g
	Saxony	1.031 lb	467.48 g
	Württemberg	1.102 lb	500.00 g
pfundschwer	Bremen	329.57 lb	149,493.00 g
pond	Netherlands	2.205 lb	1,000.00 g
pood (pud, poud)	Russia	36.07 lb	16,360.00 g
pund	Denmark	1.102 lb	500.00 g
(apothecary)	Denmark	0.789 lb	357.89 g
(apothecary)	Norway	0.786 lb	356.35 g
	Sweden	0.934 lb	423.57 g
	Norway	0.937 lb	425.16 g
quentchen	Austria	67.51 gr	4.37 g
(legal)	Bavaria	60.28 gr	3.91 g
(commercial)	Bavaria	67.51 gr	4.37 g
	Bremen	60.10 gr	3.89 g
	Brunswick	56.36 gr	3.65 g
	Darmstadt	60.28 gr	3.91 g
	Frankfurt	60.85 gr	3.95 g
	Hamburg	58.40 gr	3.78 g
	Hanover	59.02 gr	3.80 g
	Kassel	58.38 gr	3.78 g
	Lübeck	58.45 gr	3.79 g
	Prussia	60.28 gr	3.91 g
quentlein	Saxony	56.70 gr	3.61 g
quint	Denmark	1.763 oz	50.00 g
quintal	Portugal	129.52 lb	58,749.00 g
	Spain	101.61 lb	46,089.00 g
quintin	Denmark	56.32 gr	3.65 g
	Norway (ordinary)	51.25 gr	3.32 g
	(gold and silver)	50.79 gr	3.29 g
rotolo	Crete	1.166 lb	528.76 g
	Cyprus	5.242 lb	2,377.73 g
	Genoa	1.050 lb	476.42 g
	Naples	1.964 lb	891.10 g
(grosso)	Sicily	1.925 lb	872.18 g
(sottile)	Sicily	1.750 lb	793.80 g

Weights (continued)

Measure	Location	English Equivalent	Metric Equivalent
rubbio	Parma	17.99 lb	8,163.50 g
	Venetian Lombardy	22.05 lb	10,000.00 g
saum	Austria	339.54 lb	154,014.00 g
schifflast	Austria	2,469.23 lb	1,120,043.00 g
schiffpfund	Bremen	318.73 lb	144,574.00 g
	Brunswick	288.57 lb	130,894.00 g
	Hamburg	299.01 lb	135,630.00 g
scrupule	Russia	19.20 gr	1.24 g
scrupulo	Portugal	18.39 gr	1.19 g
skälpund	Norway	0.937 lb	425.16 g
	Sweden	0.934 lb	423.57 g
skippund	Norway	374.91 lb	170,061.00 g
skrupel	Austria	22.88 gr	1.48 g
	Germany	19.18 gr	1.24 g
	Norway	19.02 gr	1.23 g
stein	Austria	24.69 lb	11,201.20 g
	Baden	11.02 lb	5,000.00 g
talanton	Greece	3.307 lb	1,500.00 g
talento	Ionian Islands	100.00 lb	45,360.00 g
tomin	Spain	9.26 gr	0.60 g
tonelada	Portugal	1,748.49 lb	793,116.00 g
	Spain	2,032.19 lb	921,803.00 g
trapeso	Naples	13.75 gr	0.89 g
untz	Norway (ordinary)	0.937 oz	26.57 g
	(apothecary)	1.048 oz	29.70 g
	(gold and silver)	0.929 oz	26.33 g
untzia	Bulgaria	1.058 oz	30.00 g
unze	Austria (ordinary)	1.235 oz	35.00 g
	(apothecary)	1.238 oz	35.09 g
	Bavaria	1.235 oz	35.00 g
	Bremen	1.099 oz	31.16 g
(apothecary)	Denmark	1.052 oz	29.82 g
	Frankfurt	1.114 oz	31.58 g
(apothecary)	Germany	1.052 oz	29.82 g
	Hamburg	1.068 oz	30.28 g
	Hanover	1.079 oz	30.60 g
	Kassel	1.067 oz	30.26 g
	Lübeck	1.069 oz	30.30 g
	Saxony	1.031 oz	29.22 g
vagon	Yugoslavia	9.84 t	10.00 t
vierling	Austria	4.938 oz	140.01 g
vierding	Bavaria	4.409 oz	125.00 g

Pre-Metric European Weights and Measures

Weights (continued)

Measure	Location	English Equivalent	Metric Equivalent
wigtje	Netherlands	15.43 gr	1.00 g
zehnling	Baden	1.764 oz	50.00 g
zent	Prussia	25.72 gr	1.67 g
zentas	Baden	77.16 gr	5.00 g
zentner	Austria	123.46 lb	56,002.00 g
	Baden	110.23 lb	50,000.00 g
	Bremen	127.49 lb	57,830.00 g
	Darmstadt	110.23 lb	50,000.00 g
	Denmark	110.23 lb	50,000.00 g
	Poland	100.00 lb	45,360.00 g
	Prague	136.08 lb	61,726.00 g
	Prussia	110.23 lb	50,000.00 g
	Saxony	113.37 lb	51,423.00 g
	Sweden	93.38 lb	42,357.00 g
	Switzerland	110.23 lb	50,000.00 g
zollzentner	Germany	110.23 lb	50,000.00 g

Bibliography

Besides the rather large number of monographic and other types of studies on measurement units and physical standards, sources in other fields are relevant to one or more aspects of historical metrology. To most accurately reflect this interdisciplinary aspect, I have divided the entries in this bibliography into six major categories:

Manuscripts
Weights and Measures Monographs
Weights and Measures Articles
Documents
Reference Works
Other Works

All of the items listed in the Manuscripts section are from the British Museum Manuscripts Collection. The Weights and Measures Monographs section contains works devoted exclusively to various facets of historical metrology, including numismatics, metrological law, measuring and weighing instruments, and metric-decimal, duodecimal, binary, and other reform schemes. No documents are included in this section or in the next, Weights and Measures Articles. Primary sources appear under Documents, which include acts, assizes, decrees, ordinances, regulations, and statutes; annals; brokage and port books; calendars; cartularies; charters; chronicles; custumals; government accounts, reports, and surveys; industrial ordinances and regulations; inventories; ledgers; letters; private papers; registers; rolls (account, charter, close, court, fabric, fine, justiciary, liberate, manorial, memoranda, parliamentary, patent, pipe, and plea); university statutes; and wills. Reference Works are biographies, bibliographies, dictionaries, encyclopedias, glossaries, indexes, lexicons, manuals, tables, and word lists, all of which contain occasional definitions or discussions of weights and measures units and related metrological topics. The last section, Other Works, lists secondary sources in non-metrological fields that contain information on some aspect of weights and measures. The fields covered are law; courts and court officials; government; political-diplomatic history; church and monastery; manorialism, husbandry, and horticulture; trade, commerce, and industry; science and technology; and economic and social history.

An addendum lists recent British and American publications concerned with adoption of the metric system.

Bibliography

Manuscripts

British Museum Manuscript Collections

Add. 6159. "Register of Christchurch, Canterbury." Folios 148-148v. (Fourteenth century.)

Add. 6666. "Derbyshire Collections Analecta." Folio 299. (Seventeenth century.)

Add. 14252. Ranulphi de Glanville. "Tractatus." Folio 118v. (Twelfth century.)

Add. 17512. "S. Gregorii Dialogorum Libri." Folios 106-106v. (Eleventh century.)

Add. 32085. "Statuta tractatus varii registrum brevium." Folios 150v-151. (Fourteenth century.)

Add. 36542. "Stafford Family Evidences." Folios 5v, 7, 159-159v. (Sixteenth century.)

Calig. A. XV. "Calendar." Folios 107v-108. (Eleventh century.)

Claudius D. II. "Judicium pillorie, etc." Folios 255v-256. (Fourteenth century.)

Cotton Cleo. A. III. "Glossaries of Latin and Anglo-Saxon." Folio 10. (Eleventh century.)

Cotton Tiberius E. IV. "Annales de Winchcombe Bede de temporibus." Folio 135. (Eleventh century.)

Cotton Vesp. B. VI. "Bede de compoto." Folios 105-109. (Eighth century.)

Egerton 1925. "Traite des monnaies anciennes." (Sixteenth century.)

Galba E. IV. "Composition of Weights and Measures." Folio 29. (Fourteenth century.)

Hargrave 313. "De Scarrario." Folio 95. (Fourteenth century.)

Harley 13. "Astronomical Treatises." Folios 132v-134. (Fourteenth century.)

Harley 660. "Papers on Coins, etc." Folios 70-70v. (Seventeenth century.)

Harley 921. "Phrasæolog Latina." Folio 1. (Seventeenth century.)

Harley 1033. "Ancient Statutes, etc." Folios 135v-136. ("Modus amensurandi terram.")

Harley 1712. "Pierr Comestor Sermons." Folios 162-163. (Twelfth century.)

Harley 3205. "Assize of Inch, Ell, Perch, etc." Folios 1v-2v. (Fifteenth century.)

Harley 5394. "Eeda de figuris verborum." Folios 60v-61v. (Fourteenth century.)

Lansdowne 48. "Burghley Papers 1586." Folio 142. (Star Chamber case concerning Assize of Bread.)

Lansdowne 52. "Burghley Papers 1587." Folio 16. (Comparative table of English and classical units.)

Otho. E. X. "Papers Relating to Mines, Coinage, Weights and Measures." Folios 14-18v. (Sixteenth century.)

Royal 2 B. V. "Psalterium Cantica, etc." Folios 188-189. (Anglo-Saxon.)

Royal 7 B. X. "Johan Borough. Pupilia Occult, etc." Folios 251-251v. (Fifteenth century.)

Royal 7 DXXV. "Chronological and Other Collections." Folios 44-46. (Seventh century.)

Royal 18 CXIV. "Irish Accompts 1495-1496." Folios 154-157. (Composition of weights and measures.)

Sloane 513. "Tract on Weights and Measures by a Monk of Buckfast Abbey." Folio 25v. (Fifteenth century.)

Sloane 747. "Regist. Cartar. Abb: De Missenden." Folio 52v (Fifteenth century.)

Sloane 904. "Politica and Other Tracts." Folios 212-213. (Seventeenth century.)

Vitellus F. XII. "Collectanea." Folios 182-183. (Seventeenth century.)

Weights and Measures Monographs

Agricola, Georgi. *Medici libri quinque de mensuris et ponderibus.* Basil, 1533. (Agricola concentrates almost exclusively on the Greek and Roman systems of weights and measures and pays very little attention to the medieval.)

Alexander, J. H. *Universal Dictionary of Weights and Measures.* Baltimore, 1850. (This is one of the best compilations of weights and measures to appear in the nineteenth century. Not only does Alexander provide detailed descriptions of individual units, but he defines the standards upon which they were based.)

Arbuthnot, John. *Tables of Ancient Coins, Weights and Measures, Explain'd and Exemplify'd in Several Dissertations.* London, 1727. (Tables of weights and measures follow p. 327.)

Avery, W., and Avery, T. *Suggestions for the Amendment of the Law Relating to Weights & Measures.* London, 1888.

Bartlett-Amati, L. *Weights, Measures, Moneys and Interest Tables.* 6th ed. Rome, 1891.

Bedwell, William. *Mesolabivm architectonicvm, That Is, A Most Rare, and Singular Instrument, for the Easie, Speedy, and Most Certaine Measuring of Plaines and Solids by the Foote.* London, 1639.

Beilby, John. *Several Useful and Necessary Tables for the Gauging of Casks.* London, 1694.

Benese, Rycharde. *This Boke Sheweth the Maner of Measurynge of All Maner of Lande.* Southwarke, 1537. (Benese defines those measures that pertain to land.)

Bernardi, Edvardi. *De mensuris et ponderibus antiquis.* Oxford, 1688. (Bernardi discusses the weights and measures of Greece, Rome, and medieval England in great detail. Some of his computations, however, are incorrect.)

Berriman, A. E. *Historical Metrology.* London, 1953. (Berriman discusses the various pound weights used in medieval England and the standards for the gallon found at the Exchequer. He also comments on the historical importance of seals used in authenticating local and state standards.)

Beverini, Bartholomæo. *Syntagma de ponderibus et mensuris.* Lucca, 1711.

Blind, August. *Mass-, Münz- und Gewichtswesen*. Leipzig, 1906. (Blind describes rather superficially several medieval English units.)

Bowring, John. *The Decimal System in Numbers, Coins, and Accounts*. London, 1854.

Breed, W. Roger. *The Weights and Measures Act: 1963*. London, 1964.

Brisson, Mathurin. *Réduction des mesures et poids anciens en mesures et poids nouveaux*. Paris, 1798. (Metric conversions.)

Brown & Jackson. *The British Calculator*. London, 1814. (Contains tables and conversion factors of common British weights and measures.)

Browne, W. A. *The Money, Weights and Measures of the Chief Commercial Nations in the World with the British Equivalents*. London, 1899. (Browne treats rather superficially the systems of metrology in use before the nineteenth century.)

Chaney, H. J. *Our Weights and Measures: A Practical Treatise on the Standard Weights and Measures in Use in the British Empire with Some Account of the Metric System*. London, 1897. (Chaney outlines some of the duties of the clerks of the market in addition to defining very briefly several medieval English weights and measures.)

Clarke, Frank Wigglesworth. *Weights, Measures and Money of All Nations*. New York, 1888. (Clarke lists the weights and measures of the nineteenth century individually and by country, and he includes the location and United States-English equivalent for each unit.)

Coggeshall, Henry. *Timber Measure by a Line*. London, 1677.

———. *A Treatise of Measures*. London, 1682.

Colles, George W. *The Metric Versus the Duodecimal System*. Boston, 1896. (Originally a paper presented to the American Society of Mechanical Engineers.)

Crüger, Carl. *Contorist. Eine Handels- Münz- Mass- und Gewichtskunde*. Hamburg, 1830.

Cumberland, Richard. *An Essay towards the Recovery of the Jewish Measures and Weights, Comprehending Their Monies; by Help of Ancient Standards, Compared with Ours of England*. London, 1686.

Delambre, Jean Baptiste Joseph. *Grandeur et figure de la terre*. Ed. G. Bigourdan. Paris, 1912.

———. *Grundlagen des dezimalen metrischen Systems oder Messung des Meridianbogens zwischen den Breiten von Dünkirchen und Barcelona*. Ed. Walter Block. Leipzig, 1911.

A Digest of "The Metric Versus the English System of Weights and Measures" from Research Report No. 42. National Industrial Conference Board, Special Report No. 20. New York, 1921.

Donisthorpe, Wordsworth. *A System of Measures of Length, Area, Bulk, Weight, Value, Force, &c*. London, 1895.

The Economist Guide to Weights & Measures. London, 1956. (Compiled by the Statistical Department of *The Economist*.)

Fauve, Adrien. *Les Origines du système métrique*. Paris, 1931.

Granger, Allan. *Our Weights and Measures*. London, 1917.
Greaves, John. *A Discourse of the Roman Foot and Denarius*. London, 1737.
Guyot, Arnold. *Tables, Meterorological and Physical*. 4th ed. Smithsonian Miscellaneous Collections. Washington, D.C., 1884.
Hallock, William. *Outlines of the Evolution of Weights and Measures and the Metric System*. New York, 1906. (Among the subjects discussed by Hallock are early standards; primary and defined standards; the metrological systems of the Babylonians, Egyptians, Greeks, Romans, and Moslems; Anglo-Saxon influences; and medieval weights and measures. His remarks on medieval weights and measures are rather brief, and he tends to exaggerate the influence that ancient systems of metrology had on the development of medieval English and French units.)
Hardwicke, Robert E. *The Oilman's Barrel*. Norman, Oklahoma, 1958. (Pages 3–46 contain excellent discussions of capacity measures and their historical evolution.)
Hartmann, Carl. D*Die Waagen und ihre Construction*. Weimar, 1856.
Hauy, René Just Abbé. *Instruction sur les mesures déduites de la grandeur de la terre, uniformes pur toute la république, et sur les calculs relatifs à leur division décimale*. Paris, 1795.
Hawney, William. *The Complete Measurer; or, The Whole Art of Measuring*. London, 1789. (This is basically an arithmetic book that contains occasional descriptions of linear and superficial measures to be used for the solution of the various problems.)
Hogg, John. *Answer to a Small Treatise Call'd Just Measures*. London, 1693.
Hultsch, Fridericus. *Metrologicorum scriptorum reliquiæ*. 2 vols. Leipzig, 1864. (Hultsch includes the writings of Isidore of Seville on weights and measures. His book is especially valuable for late Roman and early medieval tracts dealing with linear measures.)
Huntar, Alexander. *A Treatise of Weights, Mets and Measures of Scotland*. Edinburgh, 1624.
Ingalls, Walter Renton. *Systems of Weights and Measures*. New York, 1945.
Jackson, Lewis D'A. *Modern Metrology: A Manual of the Metrical Units and Systems of the Present Century*. London, 1882.
Jessop, William H. R. *A Complete Decimal System of Money and Measures*. Cambridge, 1855.
Jolly, Alexander. *Conversion of Weights Tables Showing the Live & Dead Weight of Cattle, Sheep, & Pigs*. London, 1888.
Jones, Stacy V. *Weights and Measures: An Informal Guide*. Washington, D.C., 1963. (Jones makes relatively few references to medieval English weights and measures.)
Justice, Alexander. *A General Treatise of Monies and Exchanges*. London, 1707.
Keith, George Skene. *Different Methods of Establishing an Uniformity of Weights and Measures*. London, 1817.
———. *Tracts on Weights, Measures, and Coins*. London, 1791.

Kelly, P. *Metrology; or, An Exposition of Weights and Measures, Chiefly Those of Great Britain and France: Comprising Tables of Comparison, and Views of Various Standards; with an Account of Laws and Local Customs, Parliamentary Reports, & Other Important Documents.* London, 1816.

Kennelly, Arthur E. *Vestiges of Pre-Metric Weights and Measures Persisting in Metric-System Europe.* New York, 1928. (Kennelly is interested principally in French weights and measures.)

Kisch, Bruno. *Scales and Weights: A Historical Outline.* New Haven, 1965. (Kisch concentrates chiefly on the ancient and modern periods. His most important contributions are his excellent descriptions of scales and his index of important weights used in the world today.)

Klimpert, Richard. *Lexicon der Münzen, Mässe, Gewichte: Zählarten und Zeitgrössen aller Länder der Erde.* Berlin, 1896. (Klimpert is concerned primarily with French and German metrology.)

Leake, Stephen Martin. *An Historical Account of English Money from the Conquest to the Present Time.* London, 1793. (Leake discusses the troy and tower systems of weight.)

Mann, W. Wilberforce. *A New System of Measures, Weights, and Money; Entitled the Linn-Base Decimal System.* New York, 1871.

Martin, William. *An Attempt to Establish throughout His Majesty's Dominions an Universal Weight and Measure, Dependant on Each Other, and Capable of Being Applied to Every Necessary Purpose Whatever.* London, 1794.

The Metric Versus the English System of Weights and Measures. National Industrial Conference Board, Research Report No. 42. New York, 1921.

More, Richard. *The Carpenters Rule to Measure Ordinarie Timber, etc.* London, 1602.

Naft, Stephen. *Conversion Equivalents in International Trade.* Philadelphia, 1931.

Nicholson, Edward. *Men and Measures: A History of Weights and Measures: Ancient and Modern.* London, 1912. (There are tables of some medieval English, Irish, Scots, and Welsh weights and measures as well as some valuable discussions dealing with the ancient systems of metrology. Unfortunately, Nicholson exaggerates the influence that many of the ancient systems had on English metrological development. In addition, he seldom indicates the sources of his data.)

O'Keefe, John A. *The Law of Weights and Measures.* London, 1966.

———. *The Law of Weights and Measures: Supplement.* London, 1967.

Old and Scarce Tracts on Money, ed. J. R. McCulloch. London, 1933.

Oldberg, Oscar. *A Manual of Weights and Measures.* Chicago, n.d.

Oldfield, Thomas. *A Table of Silver Weight.* London, 1696.

Owen, George A. *A Treatise on Weighing Machines.* London, 1922.

Palethorpe, Joseph. *A Commercial Dictionary of the Names of All the Coins, Weights and Measures in the World.* Derby, 1829. (The title is an exaggeration.)

Monographs

Pasley, C. W. *Observations on the Expediency and Practicability of Simplifying and Improving the Measures, Weights and Money, Used in This Country, without Materially Altering the Present Standards.* London, 1834.

Paucton, Alexis Jean Pierre. *Métrologie, ou traité des mesures, poids et monnoies des anciens peuples & des modernes.* Paris, 1780.

Penkethman, John. *A Perfect Table, Declaring the Assize or Weight of Bread.* London, 1640.

Perkin, F. Mollwo. *The Metric and British Systems of Weights, Measures and Coinage.* New York, 1907.

Perry, John. *The Story of Standards.* New York, 1955.

Petrie, Sir William M. Flinders. *Measures and Weights.* London, 1934.

Records of the Coinage of Scotland from the Earliest Period to the Union, ed. R. W. Cochran-Patrick. 2 vols. Edinburgh, 1876. (There are several remarks on the avoirdupois and troy pounds.)

Reynardson, Samuel. *A State of the English Weights and Measures of Capacity.* London, 1750.

Ridgeway, William. *The Origins of Metallic Currency and Weight Standards.* Cambridge, 1892.

Romé de L'Isle, Jean Baptiste. *Métrologie, ou tables pour servir à l'intelligence des poids et mesures des anciens.* Paris, 1789.

Ruding, Rogers. *Annals of the Coinage of Great Britain and Its Dependencies; from the Earliest Period of Authentic History to the Reign of Victoria.* 2 vols. London, 1840.

Scott, George W. *The Decimal System of Weights and Measures as Authorized by Act of Congress.* Albany, 1867.

Seebohm, Frederic. *Customary Acres and Their Historical Importance.* London, 1914.

Sheppard, W. *Of the Office of the Clerk of the Market, of Weights and Measures, and of the Laws of Provision for Man and Beast.* London, 1665. (Sheppard's account of the responsibilities and duties of the clerk of the market is detailed and based on the statutes and ordinances and on his own personal observations.)

Skinner, Frederick George. *Weights and Measures: Their Ancient Origins and Their Development in Great Britain up to AD 1855.* London, 1967.

Speed, William. *Tables for Ascertaining the Weight of Cattle, Calves, Sheep, and Hogs, by Measure.* London, 1847.

Steele, John. *The Hay and Straw Measurer.* London, 1882.

Strachan, James. *A New Set of Tables for Computing the Weight of Cattle by Measurement.* London, 1849. (I also used the 1843 edition.)

Swinton, John. *A Proposal for Uniformity of Weights and Measures in Scotland, by Execution of the Laws Now in Force.* Edinburgh, 1779. (Pages 23-130 contain extremely valuable tables of the weights and measures common in the various Scottish shires.)

Tarbé, M. *Nouveau Manuel complet des poids et mesures.* Paris, 1845.

Taylor, Henry. *The Decimal System, as Applied to the Coinage & Weights & Measures of Great Britain.* 4th ed. London, 1851.

Thomson, John. *New and Correct Tables Shewing, Both in Scots and Sterling Money, the Price of Any Quantity of Grain.* Edinburgh, 1761.
Triulzi, Antonio Maria. *Bilancio dei pesi e misure di tutte le piazze mercantili dell'Europa.* 5th ed. Venice, 1803.
Vaughan, Rice. *A Discourse of Coin and Coinage.* London, 1675.
Wagstaff, W. H. *The Metric System of Weights and Measures Compared with the Imperial System.* London, 1896.
Warren, Charles. *The Ancient Cubit and Our Weights and Measures.* London, 1903.
Wurtele, Arthur. *Tables for Reducing English, Old French and Metrical Measures.* Montreal, 1861.

Weights and Measures Articles

Airy, George B. "Account of the Construction of the New National Standard of Length, and of Its Principal Copies." *Philosophical Transactions,* 147 (1857): 621–702.
Airy, Wilfrid. "On the Origin of the British Measures of Capacity, Weight and Length." *Minutes of Proceedings of the Institution of Civil Engineers,* 175 (1909): 164–76. (Airy discusses the origins of the pint, foot, and avoirdupois pound. The appendix includes some drawings of Roman and Egyptian measures together with several extracts from the laws of William I concerning standardization.)
"An Account of a Comparison Lately Made by Some Gentlemen of the Royal Society, of the Standard of a Yard, and the Several Weights Lately Made for Their Use, etc." *Philosophical Transactions,* 42 (1742–43): 544–56. (This article is concerned with the yard and ell, and with the troy and avoirdupois weight standards.)
"An Account of the Proportions of the English and French Measures and Weights, from the Standards of the Same, Kept at the Royal Society." *Philosophical Transactions,* 42 (1742–43): 185–88. (This article dwells on the similarities among the following weights and measures: Paris half-toise and English yard, Paris dimark and English troy pound, Paris foot and English foot.)
Barlow, William. "An Account of the Analogy betwixt English Weights and Measures of Capacity." *Philosophical Transactions,* 41 (1740): 457–59.
Brashaer, John A. "Evolution of Standard Measurements." *American Manufacturer and Iron World,* March 29, 1900, pp. 256–58.
Chisholm, H. W. "On the Science of Weighing and Measuring, and the Standards of Weight and Measure." *Nature,* 8 (1873): 1–192. (Two discussions by Chisholm are of special value: the toise de Pérou and the imperial standard gallon and bushel.)
Clarke, A. R. "Results of the Comparisons of the Standards of Length of England, Austria, Spain, United States, Cape of Good Hope, and of a Second Russian Standard, Made at the Ordnance Survey Office, Southampton." *Philosophical Transactions,* 163 (1873): 445–69.

Glazebrook, Sir Richard. "Standards of Measurement: Their History and Development." *Nature,* 128 (1931): 17–28. (Glazebrook concentrates on metrological standardization under Elizabeth.)

Gore, J. Howard. "The Decimal System of Measures of the Seventeenth Century." *American Journal of Science,* 41 (1891): 241–46.

Grote, George. "On Ancient Weights, Coins, and Measures." In *The Minor Works of George Grote,* ed. Alexander Bain, pp. 135–74. London, 1873.

Guilhiermoz, P. "Note sur les poids du moyen âge." *Bibliothèque de l'Ecole des Chartres,* 67 (1906): 161–233, 402–50. (Guilhiermoz discusses the weights used in most European countries, and he includes tables comparing the various pounds, which he also converts in Paris grains and metric grams.)

———. "Remarques diverses sur les poids et mesures du moyen âge." *Bibliothèque de l'Ecole des Chartres,* 80 (1919): 5–100. (The focus is again European-wide.)

Hardwicke, W. W. "Currency and Weights and Measures: The Case for Reform." *Empire Review,* December 1915, pp. 503–10.

Harkness, William. *The Progress of Science as Exemplified in the Art of Weighing and Measuring.* Smithsonian Miscellaneous Collections, Vol. 33, No. 661. Washington, D.C., 1884. (Harkness dwells on English and French measures of length before 1600, the mercantile and avoirdupois pounds, and the poids de marc and the pile de Charlemagne.)

Herschel, John F. W. "The Yard, the Pendulum, and the Metre." In *Familiar Lectures on Scientific Subjects,* pp. 419–51. New York, 1872.

Hooper, George. "An Inquiry into the State of the Ancient Measures, the Attic, the Roman, and Especially the Jewish." In *The Works of the Right Reverend George Hooper, D. D. Sometime Bishop of Bath and Wells,* Vol. 2. Oxford, 1855. (Originally published in London in 1721.)

Horwood, Alfred J. "A Manuscript in the Diocesan Library of Derry, in Ireland." In *Great Britain Historical Manuscripts Commission,* pp. 639–40. London, 1881. (Horwood describes and presents selected citations from a treatise on weights and measures, predominantly ancient, written by William Harrison sometime around 1590.)

Jefferson, Thomas. "Plan for Establishing Uniformity in the Coinage, Weights and Measures of the United States: Communicated to the House of Representatives, July 13, 1790." In *The Complete Jefferson,* ed. Saul K. Padover, pp. 974–95. New York, 1943. (Jefferson discusses many different linear, superficial, capacity, and quantity measures in addition to several types of weights.)

———. "Standards of Measures, Weights and Coins." In *The Complete Jefferson,* ed. Saul K. Padover, pp. 1004–11. New York, 1943.

Judson, Lewis Van Hagen. "Weights and Measures." *Encyclopædia Britannica,* 14th ed., rev., 1964. 23:479–88.

Kater, Henry. "An Account of the Comparison of Various British Standards of Linear Measure." *Philosophical Transactions,* 111 (1821): 75–94.

———. "An Account of the Construction and Adjustment of the New Standards of Weights and Measures of the United Kingdom of Great Britain and Ireland." *Philosophical Transactions,* 116 (1826): 1-52. (Kater lists the standards found in the cities of Edinburgh, London, and Westminster.)

———. "On the Error in Standards of Linear Measure, Arising from the Thickness of the Bar on Which They Are Traced." *Philosophical Transactions,* 120 (1830): 359-81.

McCaw, G. T. "Linear Units Old and New." *Empire Survey Review,* 5 (1939-40): 236-59. (McCaw discusses the primitive measures of length, the foot and cubit, the Greek foot, the "natural" or Olympic foot, and the Gallic leuca.)

Mendenhall, Thomas C. "Fundamental Units of Measure." In *Smithsonian Institution Annual Report,* pp. 135-49. Washington, D.C., 1893. (Originally a paper read before the International Engineering Congress of the Columbian Exposition, Chicago, 1893.)

"Metrology Universalized; or, A Proposal to Really Equalize and Universalize the Hitherto Unequalized and Arbitrary Weights and Measures of Great Britain and America." *Metric System Pamphlets,* Vol. 1, No. 8 (1828). (This article deals primarily with English linear measurement.)

Miller, Sir John Riggs. "Equalization of Weights and Measures." In *The Parliamentary History of England from the Earliest Period to the Year 1803.* Vol. 28, *1789-91,* pp. 639-50. London, 1816. (I also used the 1790 monographic version entitled *Speeches in the House of Commons upon the Equalization of the Weights and Measures of Great Britain, etc.)*

Miller, W. H. "On the Construction of the New Imperial Standard Pound, and Its Copies of Platinum; and on the Comparison of the Imperial Standard Pound with the Kilogramme des Archives." *Philosophical Transactions,* 146 (1856): 753-946. (Miller's article contains information on the tower, troy, mercantile, and avoirdupois pounds.)

Petrie, Sir William M. Flinders. "The Old English Mile." *Proceedings of the Royal Society of Edinburgh,* 12 (1882-84): 254-66.

———. "Weights and Measures." *Encyclopædia Britannica.* 14th ed., rev., 1964. 23:488H-K. (Petrie concentrates only on the ancient metrological systems.)

Prior, W. H. "Notes on the Weights and Measures of Medieval England." *Bulletin du Cange: Archivvm Latinitatis medii ævi,* 1 (1924): 77-170. (Prior defines briefly over one hundred weights and measures. He makes occasional errors in computation, but, on the whole, his work is well done.)

Rogers, W. A. "On the Present State of the Question of Standards of Length." *Proceedings of the American Academy of Arts and Sciences,* 15 (1879-80): 273-312.

Round, J. H. "Notes on Domesday Measures of Land." In *Domesday Studies,* ed. P. Edward Dove, 1: 189-225. London, 1888.

Salignacus, Bernardus. "Hides and Virgates in Sussex." *English Historical Review,* 19 (1904): 92-96.
Tait, J. "Hides and Virgates at Battle Abbey." *English Historical Review,* 18 (1903): 705-8.
"Third Report of Standards Commission, February 1, 1870." *Metric System Pamphlets,* Vol. 1, No. 2 (1878). (This short article discusses the troy and tower pounds.)
Townsend, John R. "Metric Versus English Systems." In *Systems of Units: National and International Aspects,* ed. Carl F. Kayan. Publication No. 57 of the American Association for the Advancement of Science. Washington, D.C., 1959. (Paper presented at a symposium on engineering in Washington, D.C., December 29-30, 1958.)
Vinogradoff, P. "Sulung and Hide." *English Historical Review,* 19 (1904): 282-86.
Yates, James. "Narrative of the Origin and Formation of the International Association for Obtaining a Uniform Decimal System of Measures, Weights and Coins." *Metric System Pamphlets,* Vol. 1, No. 11 (1856).
Zupko, Ronald Edward. "Medieval English Weights and Measures: Variation and Standardization." *Studies in Medieval Culture,* 4 (1974): 238-43.
———. "Notes on Medieval English Weights and Measures in Francesco Balducci Pegolotti's 'La Pratica Della Mercatura.'" *Explorations in Economic History,* 7 (1969): 153-60.
———. "The Weights and Measures of Scotland Before the Act of Union." *The Scottish Historical Review,* forthcoming.

Documents

"Account Roll of a Fifteenth-Century Iron Master," ed. G. T. Lapsley. *English Historical Review,* 14 (1899): 509-29.
Accounts of the Obedientiars of Abingdon Abbey, ed. R. E. G. Kirk. Camden Society Publication, New Series, Vol. 51. Westminster, 1892.
The Acts of the Parliaments of Scotland. Great Britain Record Commission Publications.
  A.D. MCXXIV-MCCCCXXIII. London, 1814.
  A.D. MCCCCXXIV-MDLXVII. London, 1814.
  A.D. MDLXVII-MDXCII. London, 1814.
  A.D. MDXCIII-MDCXXV. London, 1814.
  A.D. MDCXXV-MDCXLI. London, 1817.
  A.D. MDCXLI-MDCLXI. London, 1817.
  A.D. MDCLXI-MDCLXIX. London, 1820.
  A.D. MDCLXX-MDCLXXXVI. London, 1820.
  A.D. MDCLXXXIX-MDCXCV. London, 1822.
  A.D. MDCXCVI-MDCCI. London, 1823.
  A.D. MDCCII-MDCCVII. London, 1824.
Acts of the Parliaments of Scotland, 1424-1727. London, 1908. (Revised edition.)

Adams, John Quincy. *Report of the Secretary of State upon Weights and Measures Prepared in Obedience to a Resolution of the House of Representatives of the Fourteenth of December, 1819.* 16th Congress, 2d Session, House of Representatives Document No. 109. Washington, D.C., 1821. (This excellent report deals generally with the simplification and standardization of weights and measures in the early nineteenth century. Adams quotes freely from many English statutes, and he includes a short, but effective, history of English measuring units.)

Ancient Charters Royal and Private Prior to A.D. 1200: Part I. Pipe Roll Society. London, 1888.

Ancient Laws and Institutes of England, ed. Commissioners of the Public Records. London, 1840.

Ancient Laws and Institutes of Wales; Comprising Laws Supposed to be Enacted by Howel the Good, Modified by Subsequent Regulations under the Native Princes Prior to the Conquest by Edward the First, etc. Great Britain Record Commission Publications. London, 1841.

Annales Cambriæ, ed. John Williams ab Ithel. Rerum Britannicarum Medii Aevi Scriptores. London, 1860.

"Annales de Burton (A.D. 1004–1263)." In *Annales monastici,* ed. Henry Richards Luard. Rerum Britannicarum Medii Aevi Scriptores. London, 1864. (This manuscript is important for its table of unusual land measures.)

"Annales de Margan (A.D. 1066–1232)." In *Annales monastici,* ed. Henry Richards Luard. Rerum Britannicarum Medii Aevi Scriptores. London, 1864.

Annales monasterii S. Albani, a Johanne Amundesham, monacho, ut videtur, conscripti, (A.D. 1421–1440). Quibus præfigitur Chronicon rerum gestarum in monasterio S. Albani, (A.D. 1422–1431) a quodam auctore ignoto compilatum. 2 vols. Rerum Britannicarum Medii Aevi Scriptores. London, 1870–71.

Arnold, Richard. *Chronicle.* London, 1502.

The Assize of Bread with Sundry Good and Needful Ordinances. London, 1665. (This book contains translations into English of the most important assizes, among which are those for fuel and tile.)

Bartholomaei de Cotton. *Historia Anglicana (A.D. 449–1298).* Ed. Henry Richards Luard. Rerum Britannicarum Medii Aevi Scriptores. London, 1859. (Bartholomew mentions several of Richard I's decrees dealing with weights and measures.)

Beaumont, William. *An Account of the Rolls of the Honour of Halton.* Warrington, 1879.

Benedict of Peterborough. *The Chronicle of the Reigns of Henry II. and Richard I: A.D. 1169–1192.* Ed. William Stubbs. 2 vols. Rerum Britannicarum Medii Aevi Scriptores. London, 1867.

Bishop Hatfield's Survey: A Record of the Possessions of the See of Durham, Made by Order of Thomas de Hatfield, Bishop of Durham, ed. Rev. William Greenwell. Surtees Society Publication, Vol. 32. Durham, 1857.

Board of Trade. *Parliamentary Papers,* Great Britain: *Report of the Committee on Weights & Measures Legislation.* London, 1951. (Presented by the President of the Board of Trade to Parliament by command of King George VI.)

British Borough Charters: 1216-1307, ed. A. Ballard and J. Tait. Cambridge, 1923.

The Brokage Book of Southampton: 1443-1444, ed. Olive Coleman. 2 vols. Southampton, 1960-61. (Volume 2 is valuable for its lists of products.)

Calendarium rotulorum patentium in turri Londinensi. London, 1802.

Calendar of County Court, City Court and Eyre Rolls of Chester, 1259-1297, ed. R. Stewart-Brown. Chetham Society Publication, Vol. 84. Manchester, 1925.

Calendar of the Charter Rolls Preserved in the Public Record Office. (Most of the volumes of the various Rolls included in my Bibliography contain information dealing with infractions of statutory standards and with the punishments imposed for violations. Seldom are there actual tables of individual units.)

Henry III: 1216-1257. London, 1904.
Henry III-Edward I: 1257-1300. London, 1906.
Edward I-Edward II: 1300-1326. London, 1908.
15 Edward III-5 Henry V: 1341-1417. London, 1916.
5 Henry VI-8 Henry VIII: 1427-1516. London, 1927.

Calendar of the Close Rolls Preserved in the Public Record Office.

Edward II: 1302-1307. London, 1908.
Edward II: 1313-1318. London, 1893.
Edward II: 1318-1323. London, 1895.
Edward II: 1323-1327. London, 1898.
Edward III: 1330-1333. London, 1898.
Edward III: 1339-1341. London, 1901.
Edward III: 1343-1346. London, 1904.
Edward III: 1349-1354. London, 1906.
Edward III: 1354-1360. London, 1908.
Edward III: 1360-1364. London, 1909.
Edward III: 1369-1374. London, 1911.
Edward III: 1374-1377. London, 1911.
Richard II: 1377-1381. London, 1914.
Richard II: 1381-1385. London, 1920.
Richard II: 1392-1396. London, 1925.
Richard II: 1396-1399. London, 1927.
Henry IV: 1399-1402. London, 1927.
Henry IV: 1402-1405. London, 1929.
Henry IV: 1405-1409. London, 1931.
Henry IV: 1409-1413. London, 1932.
Henry V: 1413-1419. London, 1929.
Henry V: 1419-1422. London, 1932.
Henry VI: 1435-1441. London, 1937.

Henry VI: 1441–1447. London, 1937.
Henry VI: 1447–1454. London, 1947.
Henry VII: 1500–1509. London, 1963.
Calendar of the Fine Rolls Preserved in the Public Record Office.
 Edward I: 1272–1307. London, 1911.
 Edward II: 1319–1327. London, 1912.
 Edward III: 1337–1347. London, 1915.
Calendar of the Justiciary Rolls or Proceedings in the Court of the Justiciar of Ireland, ed. James Mills. 2 vols. London, 1914.
Calendar of the Letter Books of the City of London, ed. R. H. Sharpe. London, 1899.
Calendar of the Liberate Rolls Preserved in the Public Record Office.
 Henry III: 1226–1240. London, 1916.
 Henry III: 1240–1245. London, 1930.
 Henry III: 1245–1251. London, 1937.
 Henry III: 1251–1260. London, 1959.
 Henry III: 1267–1272. London, 1959.
Calendar of the Patent Rolls Preserved in the Public Record Office.
Calendar of the Patent Rolls Preserved in the Public Record Office.
 Henry III: 1266–1272. London, 1913.
 Edward I: 1272–1281. London, 1901.
 Edward II: 1307–1313. London, 1894.
 Edward II: 1317–1321. London, 1903.
 Edward III: 1327–1330. London, 1891.
 Edward III: 1338–1340. London, 1898.
 Edward III: 1340–1343. London, 1900.
 Edward III: 1343–1345. London, 1902.
 Edward III: 1345–1348. London, 1903.
 Edward III: 1348–1350. London, 1905.
 Edward III: 1350–1354. London, 1907.
 Edward III: 1354–1358. London, 1909.
 Edward III: 1358–1361. London, 1911.
 Edward III: 1364–1367. London, 1912.
 Henry IV: 1399–1401. London, 1903.
 Henry VI: 1422–1429. Norwich, 1901.
 Edward IV: 1461–1467. London, 1897.
 Edward IV, Edward V, and Richard III: 1476–1485. London, 1901.
 Edward VI: 1547–1548. London, 1924.
 Edward VI: 1548–1549. London, 1924.
 Edward VI: 1549–1551. London, 1925.
 Edward VI: 1547–1553. London, 1937.
 Edward VI: 1550–1553. London, 1926.
 Philip and Mary: 1553–1554. London, 1937.
Calendar of the Plea and Memoranda Rolls, ed. A. H. Thomas. Cambridge, 1929.

Capgrave, John. *The Chronicle of England.* Ed. Rev. Francis Charles Hingeston. Rerum Britannicarum Medii Aevi Scriptores. London, 1858.
Cartularium Monasterii de Rameseia, ed. William Henry Hart and Rev. Ponsonby A. Lyons. Rerum Britannicarum Medii Aevi Scriptores. London, 1884-93.
Cartulary of St. Mary Clerkenwell, ed. W. O. Hassall. Camden Third Series, Vol. 71. London, 1949.
The Cely Papers: Selections from the Correspondence and Memoranda of the Cely Family, Merchants of the Staple, A.D. 1475-1488, ed. Henry Elliot Malden. Camden Society Publication, Third Series, Vol. 1. London, 1900.
Charters and Documents Illustrating the History of the Cathedral, City, and Diocese of Salisbury, in the Twelfth and Thirteenth Centuries, ed. Rev. W. Dunn Macray. Rerum Britannicarum Medii Aevi Scriptores. London, 1891.
The Charters of Endowment, Inventories, and Account Rolls, of the Priory of Finchale, in the County of Durham. Surtees Society Publications, Vol. 6. London, 1837.
Charters of the Cinque Ports, Two Ancient Towns, and Their Members, ed. Samuel Jeake. London, 1728.
Chartularies of St. Mary's Abbey, Dublin: With the Register of its House at Dunbrody, and Annals of Ireland, ed. John T. Gilbert. Vol. 2. Rerum Britannicarum Medii Aevi Scriptores. London, 1884.
Chronica Johannis de Oxenedes, ed. Sir Henry Ellis. Rerum Britannicarum Medii Aevi Scriptores. London, 1859.
Chronica monasterii de Melsa, a fundatione usque ad annum 1396, auctore Thoma de Burton, abbate, ed. Edward A. Bond. 3 vols. Rerum Britannicarum Medii Aevi Scriptores. London, 1870-71.
Chronica Rogeri de Hoveden, ed. William Stubbs. Vols. 3 and 4. Rerum Britannicarum Medii Aevi Scriptores. London, 1866.
"The Chronicle of Richard of Devizes." In *Chronicles of the Reigns of Stephen, Henry II, and Richard I,* ed. Richard Howlett, 3:381-454. Rerum Britannicarum Medii Aevi Scriptores. London, 1870-71.
Chronicle of the Mayors and Sheriffs of London, ed. H. T. Riley. London, 1863.
Chronicles of London, ed. Charles Lethbridge. Oxford, 1905.
Chronicon abbatiæ de Evesham, ad annum 1418, ed. William Dunn Macray. Rerum Britannicarum Medii Aevi Scriptores. London, 1863.
Chronicon Monasterii de Abingdon, ed. Rev. Joseph Stevenson. Rerum Britannicarum Medii Aevi Scriptores. London, 1858.
Chronicon Monasterii de Bello. Anglia Christiana Society. London, 1846. (This is a particularly valuable source for superficial measures.)
Close Rolls of the Reign of Henry III Preserved in the Public Record Office.
Henry III: *1227-1234.* London, 1902.
Henry III: *1234-1237.* London, 1908.
Henry III: *1237-1242.* London, 1911.

Henry III: 1242-1247. London, 1916.
Henry III: 1247-1251. London, 1922.
Henry III: 1251-1253. London, 1927.
Henry III: 1254-1256. London, 1931.
Henry III: 1261-1264. London, 1936.
Henry III: 1268-1272. London, 1938.
Collection complète des lois, décrets, ordonnances, réglemens, avis du Conseil d'état, publiée sur les éditions officielles du Louvre; de l'Imprimerie nationale, par Boudouin; et du Bulletin des Lois; de 1788 à 1830 inclusivement, ed. J. B. Duvergier. 30 vols. Paris, 1834. (Although most of the metrological information contained in these volumes is French, there are references to English weights and measures, especially for the sake of comparison.)
Collection des décrets de l'Assemblée nationale constituante, ed. M. Arnoult. Vols. 2 and 6. Dijon, 1792. (Mainly important for French metrology, these volumes contain references to some English weights and measures.)
A Collection in English of the Statutes Now in Force, Continued from the Beginning of Magna Charta, Made in the 9. Yere of the Raigne of King H. 3. until the End of the Parliament Holden in the 7. Yere of the Raigne of Our Soveraigne Lord King James. London, 1615, (This collection provides some valuable commentaries on the statutes, and it is especially important as a source of variant spellings.)
A Collection of Acts and Ordinances of General Use, Made in the Parliament Begun and Held at Westminster the Third Day of November, Anno 1640, ed. Henry Scobell. London, 1658.
A Collection of the Statutes Made in the Reigns of King Charles the I and King Charles the II, ed. Tho. Manby. London, 1687.
A Common-place Book of the Fifteenth Century, ed. Lucy Toulmin Smith. London, 1886.
A Consuetudinary of the Fourteenth Century for the Refectory of the House of S. Swithun in Winchester, ed. George William Kitchin. London, 1886.
Continuation of the Court Leet Records of the Manor of Manchester: A.D. 1586-1602, ed. John Harland. Chetham Society Publication, Vol. 65. Manchester, 1865.
Coopers Company, London. *Historical Memoranda, Charters, Documents, and Extracts, from the Records of the Corporation and the Books of the Company, 1396-1848.* London, 1848. (Compiled by James F. Firth, a member of the company.)
Court Rolls of the Manor of Ingoldmells, ed. W. C. Massingberd. London, 1902.
The Coventry Leet Book: or Mayor's Register, Containing the Records of the City Court Leet or View of Frankpledge: 1420-1555, ed. Mary Dormer Harris. Early English Text Society. London, 1913. (This is one of the most important sources for information on capacity measures.)
Cronica Jocelini de Brakelonda, de rebus gestis Samsonis abbatis monasterii Sancti Edmundi, ed. Johanne Gage Rokewode. Camden Society Publication, Vol. 13. London, 1840.

Croniques de London, depuis l'an 44 Hen. III. jusqu'a'l'an 17 Edw. III, ed. George James Aungier. Camden Society Publication, Vol. 28. London, 1844.
Curia Regis Rolls of the Reign of Henry III Preserved in the Public Record Office.
  3-4 Henry III. London, 1938.
  4-5 Henry III. London, 1952.
  5-6 Henry III. London, 1949.
  7-9 Henry III. London, 1955.
  11-14 Henry III. London, 1959.
  14-17 Henry III. London, 1961.
Curia Regis Rolls of the Reigns of Richard I and John Preserved in the Public Record Office.
  Richard 1-2 John. London, 1922.
  3-5 John. London, 1925.
  5-7 John. London. 1926.
  7-8 John. London, 1929.
  8-10 John. London, 1931.
  11-14 John. London, 1932.
  15-16 John. London, 1935.
Custumals of Battle Abbey in the Reigns of Edward I and Edward II (1283-1312) from MSS in the Public Record Office, ed. S. R. Scargill-Bird. Camden Society Publication, New Series, Vol. 41. Westminster, 1887. (The documents provide information on the wista and other superficial measures.)
Documents Relating to the Foundation of the Chapter of Winchester: A.D. 1541-1547, ed. George Williams Kitchin and Francis Thomas Madge. London, 1889.
Domesday Book in Relation to the County of Sussex, ed. William Douglas Parish. Sussex, 1886.
The Domesday Book of Kent, ed. Rev. Lambert Blackwell Larking. London, 1869.
The Domesday of St. Paul's of the Year MCCXXII; or, Registrum de visitatione maneriorum per Robertum Decanum, ed. William Hale. Camden Society Publication, Vol. 69. Westminster, 1858. (This is a particularly important source for capacity measures and for large superficial measures such as the hide and virgate.)
Domesday Tables for the Counties of Surrey, Berkshire, Middlesex, Hertford, Buckingham and Bedford and for the New Forest, ed. Francis Henry Baring. London, 1909.
Drei volkswirtschaftliche Denkschriften aus der Zeit Heinrichs VIII. von England, ed. Reinhold Pauli. Göttingen, 1878.
The Durham Household Book; or, The Accounts of the Bursar of the Monastery of Durham from Pentecost 1530 to Pentecost 1534. Surtees Society Publication, Vol. 18. London, 1844. (The glossary contains descriptions of some capacity measures and of several types of cloth.)

Elton Manorial Records: 1279–1351, ed. S. C. Ratcliff. The Roxburghe Club. Cambridge, 1946.

English Gilds: The Original Ordinances of More Than One Hundred Early English Gilds, ed. Lucy Toulmin Smith. Early English Text Society. London, 1870. (There are occasional references to weights and measures and to the verification and enforcement of gild standards.)

The Estate Book of Henry de Bray of Harleston, Co. Northants (c. 1289–1340), ed. Dorothy Willis. Camden Third Series, Vol. 27. London, 1916. (There is a table of land measures among the documents.)

An Exact Abridgement of the Records in the Tower of London, ed. Sir Robert Cotton. London, 1657. (This collection contains few references to weights and measures, but occasionally there are summaries of the principal statutes and ordinances that dealt with them.)

Excerpta e Libris Domicilii domini Jacobi quinti regis Scotorum, ed. The Bannatyne Club. Edinburgh, 1836.

Expeditions to Prussia and the Holy Land Made by Henry Earl of Derby (afterwards King Henry IV) in the Years 1390–1 and 1392–3 Being the Accounts Kept by His Treasurer during Two Years, ed. Lucy Toulmin Smith. Camden Society Publication, New Series, Vol. 52. London, 1894. (This account of expenditures contains some valuable information on capacity measures.)

Extracts from the Records of the Company of Hostmen of Newcastle-upon-Tyne. Surtees Society Publication, Vol. 105. Durham, 1901.

Extracts from the Records of the Merchant Adventurers of Newcastle-upon-Tyne. Surtees Society Publication, Vol. 93. Durham, 1895. (There is a detailed discussion of the chalder and the keel.)

The Fabric Rolls of York Minster. Surtees Society Publication, Vol. 35. Durham, 1859. (The glossary contains definitions of several capacity and superficial measures along with documentary materials illustrating their use.)

Fabyan, Robert. *The Newe Chronycles of Englande and of Fraunce.* London, 1516.

Feet of Fines for the County of Lincoln for the Reign of King John (1199–1216), ed. Margaret S. Walker. London, 1954.

Feet of Fines for the County of Norfolk for the Reign of King John (1201–1215) and for the County of Suffolk for the Reign of King John (1199–1214), ed. Barbara Dodwell. London, 1958. (This work is important for its excellent descriptions of superficial measures such as the bovate, virgate, and knight's fee. The subject index also contains some valuable information.)

Feet of Fines for the County of Norfolk for the Tenth Year of the Reign of King Richard the First (1198–1199) and for the First Four Years of the Reign of King John (1199–1202), ed. Barbara Dodwell. London, 1952.

Feudal Documents from the Abbey of Bury St. Edmunds, ed. D. C. Douglas. London, 1932.

Fleta, ed. H. G. Richardson and G. O. Sayles. Selden Society Publication. London, 1955. (*Fleta* is a valuable source for many capacity measures and for information pertaining to the construction of the tower and mercantile pounds.)

Flores Historiarum, ed. Henry Richards Luard. 2 vols. Rerum Britannicarum Medii Aevi Scriptores. London, 1890.

The Great Charter Called I(n) Latyn Magna Carta with Divers Olde Statutes. London, 1541.

The Great Rolls of the Pipe for the Seventeenth Year of the Reign of King Henry the Second: A.D. 1170-1. Pipe Roll Society. London, 1893.

Hall, Hubert, and Nicholas, Frieda J., eds. *Select Tracts and Table Books Relating to English Weights and Measures (1100-1742).* Camden Third Series, Vol. 41. London, 1929. (This is the most complete single collection dealing specifically with medieval English weights and measures. The documents cover all five major divisions of measurement as well as cloth regulations. The Cottonian manuscripts have been used in addition to other valuable collections. With few exceptions, the editing is well done; important information is contained in the footnotes.)

Hassler, Ferdinand Rudolph. *Report upon the Comparison of Weights and Measures of Length and Capacity, Made at the City of Washington, in 1831, under the Direction of the Treasury Department, in Compliance with a Resolution of the Senate of the United States of the 29th May, 1830.* 22d Congress, 1st Session, House of Representatives Document No. 299. Washington, D.C., 1832. (There is some information dealing with the composition of English gallons and bushels.)

Heales, Alfred. *The Records of Merton Priory in the County of Surrey.* London, 1898.

Henrici de Bracton. *De legibus et consuetudinibus Angliæ*, ed. Travers Twiss. 6 vols. Rerum Britannicarum Medii Aevi Scriptores. London, 1878-83.

Hereafter Ensueth the Auncient Customes of the Mannors of the Sebbenhuth. London, 1610.

Historia et Cartularium Monasterii Sancti Petri Gloucestriæ, ed. William Henry Hart. Vol. 3. Rerum Britannicarum Medii Aevi Scriptores. London, 1867.

Historic and Municipal Documents of Ireland, A.D. 1172-1320, ed. J. T. Gilbert. Rerum Britannicarum Medii Aevi Scriptores. London, 1870.

The House and Farm Accounts of the Shuttleworths of Gawthorpe Hall, in the County of Lancaster, at Smithils and Gawthorpe, from September 1582 to October 1621, ed. John Harland. 2 vols. Chetham Society Publication, Vols. 43 and 46. London, 1858. (Both volumes contain glossaries with descriptions of the weights and measures found on this estate.)

The Inventories and Account Rolls of the Benedictine Houses or Cells, of Jarrow and Monk-Wearmouth, in the County of Durham. Surtees Society Publication, Vol. 29. Durham, 1854. (There is information on weights and measures in the glossary.)

Issues of the Exchequer: Being Payments Made Out of His Majesty's Revenue during the Reign of King James I, ed. Frederick Devon. London, 1836.

The Itinerary in Wales of John Leland in or about the Years 1536-1539, ed. Lucy Toulmin Smith. London, 1906.

The Kalendar of Abbot Samson of Bury St. Edmunds and Related Documents, ed. R. H. C. Davis. Camden Third Series, Vol. 84. London, 1954. (This book contains information on socage land and socage dues in addition to several descriptions of unusual capacity measures found at St. Edmunds.)

Langtoft, Peter. *Chronicle.* Ed. Thomas Hearne. London, 1810. (The glossary contains descriptions of the larger superficial measures such as the farthingdale, virgate, and hide.)

Laws and Acts of Parliament: James VII to Anne (1685-1707). Edinburgh, 1731.

The Laws of the Earliest English Kings, ed. F. L. Attenborough. Cambridge, 1922.

Ledger of Andrew Halyburton, Conservator of the Privileges of the Scotch Nation in the Netherlands: 1492-1503, Together with the Book of Customs and Valuation of Merchandises in Scotland: 1612. Edinburgh, 1867.

Leet Jurisdiction in the City of Norwich during the XIIIth and XIVth Centuries, ed. William Hudson. Selden Society Publication, Vol. 5. London, 1892. (Some of the cases deal with infractions of metrological regulations and with the various fines and amercements levied as penalties.)

The Letter Books of Joseph Holroyd (Cloth-Factor) and Sam Hill (Clothier): Documents Illustrating the Organisation of the Yorkshire Textile Industry in the Early 18th Century, ed. Herbert Heaton. County Borough of Halifax, Bankfield Museum Notes, Second Series, No. 3. Halifax, 1914.

"Letter from the Secretary of the Interior Transmitting in Response to a Resolution of the House of Representatives, Reports concerning the Adoption of the Metric System of Weights and Measures." *Metric System Pamphlets,* Vol. 1, No. 7 (1878).

"Letter from the Secretary of the Treasury Transmitting to the House of Representatives Certain Reports in Reference to the Adoption of the Metric System." *Metric System Pamphlets,* Vol. 1, No. 2 (1878).

"Letter from the Secretary of War Transmitting Reports of Chiefs of Bureaus upon the Adoption of the Metrical System, in Response to a Resolution of the House of Representatives." *Metric System Pamphlets,* Vol. 1, No. 6 (1878).

Liber monasterii de Hyda; Comprising a Chronicle of the Affairs of England, from the Settlement of the Saxons to the Reign of King Cnut; and a Chartulary of the Abbey of Hyde, in Hampshire: A.D. 455-1023, ed. Edward Edwards. Rerum Britannicarum Medii Aevi Scriptores. London, 1866.

Liber quotidianus contraolulatoris garderobæ: Anno regni regis Edwardi prime: A.D. MCCXCIX and MCCC. London, 1787. (There are several descriptions of heaped, striked, and shallow capacity measures together with a discussion of the crannock.)

Lingelbach, W. E. *The Merchant Adventurers of England: Their Laws and Ordinances with Other Documents.* Philadelphia, 1902.
The Local Port Book of Southampton for 1435-36, ed. Brian Foster. Southampton, 1963. (This and the port book for 1439-40 provide numerous examples of the types of capacity and quantity measures that were used for certain products. Occasionally there are descriptions of the measures themselves.)
The Local Port Book of Southampton for 1439-40, ed. Henry S. Cobb. Southampton, 1961.
The Maire of Bristowe Is Kalendar by Robert Ricart, Town Clerk of Bristol 18 Edward IV, ed. Lucy Toulmin Smith. Camden Society Publication, New Series, Vol. 5. Westminster, 1872. (There is a description of the crannock in this calendar.)
The Manor of Manydown Hampshire, ed. G. W. Kitchin. London, 1895. (There is a brief discussion of superficial measurement in the introduction.)
Matthaei Parisiensis (Matthew of Paris). *Chronica Majora.* Ed. Henry Richards Luard. Rerum Britannicarum Medii Aevi Scriptores. 6 vols. London, 1874. (Matthew refers to Richard I's attempt to standardize weights and measures in 1189 in a section entitled "De persecutione Judæorum."
———. *Historia Anglorum.* Ed. Sir Frederic Madden. Rerum Britannicarum Medii Aevi Scriptores. 3 vols. London, 1865-69.
Medieval Archives of the University of Oxford, ed. Rev. H. E. Salter. Oxford, 1921. (There are several descriptions of the seals used by the Chancellor of Oxford University in authenticating weights and measures under his jurisdiction.)
Memorials of London and London Life in the XIIIth, XIVth and XVth Centuries, ed. Henry Thomas Riley. London, 1868. (There are several descriptions of capacity measures in addition to a number of inventories. The footnotes occasionally give information on the types of scales used by merchants.)
Memorials of St. Giles, Durham, Being Grassmen's Accounts and Other Parish Records, Together with Documents Relating to the Hospital of Kepier and St. Mary Magdalene. Surtees Society Publication, Vol. 95. Durham, 1896. (This volume contains information on the standard weights for bread.)
Memorials of the Abbey of St. Mary of Fountains, ed. Joseph Thomas Fowler. Surtees Society Publication, Vol. 130. Durham, 1918. (The glossary has definitions for several dry and liquid capacity measures.)
Monumenta Franciscana: Being a Further Collection of Original Documents Respecting the Franciscan Order in England, ed. Richard Howlett. Vol. 2. Rerum Britannicarum Medii Aevi Scriptores. London, 1882.
Monumenta Juridica: The Black Book of the Admiralty, ed. Sir Travers Twiss. Rerum Britannicarum Medii Aevi Scriptores. London, 1873.
Munimenta academica, or Documents Illustrative of Academical Life and Studies at Oxford, ed. Rev. Henry Anstey. Rerum Britannicarum Medii

Aevi Scriptores. London, 1868. (The duties of the Chancellor of Oxford University in regard to the maintenance of Crown standards are described in several documents.)

Munimenta gildhallæ Londoniensis: Liber Albus, Liber Custumarum et Liber Horn, ed. Henry Thomas Riley. Rerum Britannicarum Medii Aevi Scriptores. London, 1859-62. (The *Liber Albus* and the *Liber Custumarum* are the most important for information on weights and measures.)

Murray, T., of Glendook. *Laws and Acts of Parliament Made by James I, II, III, IV, V, Queen Mary, James VI, Charles I and II (1424-1681).* Edinburgh, 1682-85.

Northumberland Household Book (The Regulations and Establishment of the Household of Henry Algernon Percy, the Fifth Earl of Northumberland) 1512-25. London, 1770.

"Palatinate of Chester: 1351-1365." Part 3 of *Register of Edward the Black Prince.* London, 1932. (Several documents are concerned with the standardization of Cheshire's weights and measures.)

The Pipe Roll of the Bishopric of Winchester for the Fourth Year of the Pontificate of Peter des Roches, 1208-1209, ed. Hubert Hall. London, 1903.

Polychronicon Ranulphi Higden monachi cestrensis, ed. Churchill Babington et al. 9 vols. Rerum Britannicarum Medii Aevi Scriptores. London, 1865-86. (The volumes also contain translations of the *Polychronicon* by John Trevisa, A.D. 1387, and by an anonymous author of the fifteenth century.)

Radulphi de Coggeshall Chronicon Anglicanum, ed. Josephus Stevenson. Rerum Britannicarum Medii Aevi Scriptores. London, 1875.

The Rates of the Custome House Bothe Inwarde and Outwarde the Difference of Measures and Weyghts and Other Commodities Very Necessarye for All Marchantes to Knowe Newly Correctyd and Imprynted. London, 1545.

The Rates of the Custome House Reduced into a Much Better Order for the Redier Finding of Any Thing Therin Contained. London, 1590.

Records of Some Salford Portmoots, ed. J. Tait. Chetham Society, New Series, Vol. 80. Manchester, 1926.

Records of the Borough of Nottingham: Being a Series of Extracts from the Archives of the Corporation of Nottingham. 2 vols. London, 1883.

Records of the City of Norwich, ed. W. Hudson and J. C. Tingey. Norwich, 1906.

Recueil général des anciennes lois françaises, depuis l'an 420 jusqu' à la révolution de 1789, ed. MM. Jourdan, Decrusy, and Isambert. 30 vols. Paris, 1830. (Information on weights and measures is contained in volumes 1, 3, 9, 11, 13, 14, 18, 20, 22, 24, 25, 26, and 27.)

The Red Book of the Exchequer, ed. Hubert Hall. 3 vols. Rerum Britannicarum Medii Aevi Scriptores. London, 1896.

Regestra regum Scottorum: The Acts of Malcolm IV King of Scots, 1153-1165, ed. G. W. S. Barrow. Edinburgh, 1960.

The Register of Malmesbury Abbey, ed. J. S. Brewer and Charles Trice Martin. 2 vols. Rerum Britannicarum Medii Aevi Scriptores. London, 1886.

The Register of S. Osmund, ed. W. H. Rich Jones. 2 vols. Rerum Britannicarum Medii Aevi Scriptores. London, 1883-84.

Registrum Honoris de Richmond, ed. Roger Gale. London, 1722. (Especially important are the discussions of land measures in the appendixes; Appendix 2, *Of Dimensions of Lands,* was compiled by Arthur Agard in 1599 during his tenure as Deputy Chamberlain of the Exchequer from 1570 to 1615.)

Registrum sive liber irrotularius et consuetudinarius prioratus Beatæ Mariæ Wigorniensis, ed. William Hale. Camden Society Publication, Vol. 91. London, 1865.

Registrum Vulgariter Nuncupatum: The Record of Caernarvon. London, 1838. (This collection of documents contains the famous description of the tower pound so often found in other medieval manuscripts. There is also a version of the Assize of Bread.)

A Relation, or Rather a True Account, of the Island of England; with Sundry Particulars of the Customs of These People, and of the Royal Revenues under King Henry the Seventh, about the Year 1500, ed. Charlotte Augusta Sneyd. Camden Society Publication, Vol. 37. London, 1847. (This collection contains several lists of products and the capacity measures by which they were sold.)

"Report from the Committee Appointed to Inquire into the Original Standards of Weights and Measures in This Kingdom, and to Consider the Laws Relating Thereto." *Parliamentary Papers,* Great Britain: *Reports from Committees of the House of Commons,* 2 (1737-65): 411-51. (This report contains some excellent discussions of medieval English weights and measures. Ample documentation, especially from *Fleta,* and concise summaries of metrological laws make this an important source.)

"Report from the Committee Appointed (upon the 1st Day of December, 1758) to Inquire into the Original Standards of Weights and Measures in This Kingdom; and to Consider the Laws Relating Thereto." *Parliamentary Papers,* Great Britain: *Reports from Committees of the House of Commons,* 2 (1737-65): 455-63.

"Report from the Select Committee of the House of Lords Appointed to Consider the Petition of the Directors of the Chamber of Commerce and Manufactures, Established by Royal Charter in the City of Glasgow Taking Notice of the Bill Entitled 'An Act for Ascertaining and Establishing Uniformity of Weights and Measures etc.' " *Parliamentary Papers,* Great Britain: *Reports from Committees of the House of Lords,* 7 (1824): 1-35.

"Report from the Select Committee on the Weights and Measures Act; Together with the Minutes of Evidence." *Parliamentary Papers,* Great Britain: *Reports from Committees of the House of Lords,* 18 (1835): 1-60.

"Report of the Committee Appointed to Superintend the Construction of the New Parliamentary Standards of Length and Weight." *Parliamentary Papers,* Great Britain: *Reports from Committees of the House of Lords,* 19 (1854): 1-23.

Rich, E. E. *The Ordinance Book of the Merchants of the Staple.* Cambridge, 1937. (The *Ordinance Book of 1565* covers pages 103-200.)

A Roll of the Household Expenses of Richard de Swinfield, Bishop of Hereford during Part of the Years 1289 and 1290, ed. Rev. John Webb. Camden Society Publication, Vol. 59. London, 1854. (The glossary contains descriptions of several weights and measures.)

Rolls of the Justices in Eyre Being the Rolls of Pleas and Assizes for Lincolnshire 1218-9 and Worcestershire 1221, ed. Doris Mary Stenton. Selden Society Publication, Vol. 53. London, 1934.

Rotuli parliamentorum Anglie hactenus inediti MCCLXXIX-MCCCLXXIII, ed. H. G. Richardson and George Sayles. Camden Third Series, Vol. 51. London, 1935.

Rotuli parliamentorum ut et petitiones, et placita in parliamento tempore, Edwardi R. I, ed. Rev. John Strachey et al. Vol. 1 (1278-1325). London, 1832. (The *Rotuli parliamentorum* are especially valuable for information on capacity measures and for descriptions of cloth measurements.)

Rotuli parliamentorum ut et petitiones, et placita in parliamento tempore Edwardi R. III, ed. Rev. John Strachey et al. Vol. 2 (1326-1377). London, 1832.

Rotuli parliamentorum ut et petitiones, et placita in parliamento tempore Ricardi R. II, ed. Rev. John Strachey et al. Vol. 3 (1377-1411). London, 1832.

Rotuli parliamentorum ut et petitiones, et placita in parliamento tempore Henrici R. V, ed. Rev. John Strachey et al. Vol. 4 (1413-1437). London, 1832.

Rotuli parliamentorum ut et petitiones, et placita in parliamento ab anno decimo octavo R. Henrici sexti ad finem ejusdem regni, ed. Rev. John Strachey et al. Vol. 5 (1439-1468). London, 1832.

Rotuli parliamentorum ut et petitiones, et placita in parliamento ab anno duodecimo R. Edward IV ad finem ejusdem regni, ed. Rev. John Strachey et al. Vol. 6 (1472-1503). London, 1832.

The Royal Charters of Grantham: 1463-1688, ed. G. H. Martin. Leicester, 1963.

Rural Economy in Yorkshire in 1641, Being the Farming and Account Books of Henry Best, of Elmswell, in the East Riding of the County of York. Surtees Society Publication, Vol. 33. Durham, 1857. (The glossary contains descriptions of the leap and maund.)

Scotland in 1298: Documents Relating to the Campaign of King Edward the First in That Year, and Especially to the Battle of Falkirk, ed. Henry Gough. London, 1888.

Scotland: The Lawes and Actes of Parliament. Edinburgh, 1597.

Scots Statutes Revised, 1424-1900. 10 vols. London, 1899-1907.

"Second Report of the Commissioners Appointed by His Majesty to Consider the Subject of Weights and Measures." *Parliamentary Papers,* Great Britain: *Reports from Commissioners,* 7 (1820): 1-40. (This report is extremely valuable, for it includes a 40-page listing of the state and local units of measurement that were common in the late 1700s and early 1800s.)

Select Cases before the King's Council in the Star Chamber Commonly Called the Court of Star Chamber, ed. I. S. Leadam. Selden Society Publications, Vol. 26. London, 1911. (This book contains information on the wey, firkin, butter barrel, and virgate.)

Select Cases concerning the Law Merchant: A.D. 1270-1638, ed. Charles Gross. Selden Society Publication, Vol. 23 (Local Courts). London, 1908. (There are several descriptions of infractions of weights and measures legislation together with a glossary, which defines such capacity measures as the trey and ring.)

Select Cases concerning the Law Merchant: A.D. 1239-1633, ed. Hubert Hall. Selden Society Publication, Vol. 46 (Central Courts). London, 1930.

Select Cases concerning the Law Merchant: A.D. 1251-1779, ed. Hubert Hall. Selden Society Publication, Vol. 49 (Central Courts). London, 1932.

Select Cases in the Council of Henry VII, ed. C. G. Bayne and William Huse Dunham, Jr. Selden Society Publication, Vol. 75. London, 1958.

"The Select Committee Appointed to Consider the Several Reports Which Have Been Laid before This House, Relating to Weights and Measures." *Parliamentary Papers,* Great Britain: *Reports from Committees,* 4 (1821): 1-7.

Selected Rolls of the Chester City Courts, ed. A. Hopkins. Chetham Society, Third Series, Vol. 2. Manchester, 1950.

Select English Historical Documents of the Ninth and Tenth Centuries, ed. F. E. Harmer. Cambridge, 1914. (There are relatively few documents dealing specifically with weights and measures.)

Select Pleas in Manorial and Other Seignorial Courts, ed. F. W. Maitland. Selden Society Publication, Vol. 28. London, 1889.

Select Pleas of the Crown, ed. F. W. Maitland. Selden Society Publication, Vol. 27. London, 1888.

"Seventh Annual Report of the Warden of the Standards on the Proceedings and Business of the Standard Weights and Measures Department of the Board of Trade for 1872-73." *Parliamentary Papers,* Great Britain: *Reports from Commissioners,* 38 (1873): 1-105. (This long report deals primarily with the laws relating to the inspection and verification of weights and measures. There are some excellent metrological tables in addition to several photographs of common capacity measures.)

Silegrave, Henry of. *A Chronicle of English History from the Earliest Period to A.D. 1274,* Ed. C. Hock. London, 1849. (Printed for the Caxton Society and published from a Cottonian MS.)

Skene, Sir John. *Regiam majestatum, the Auld Lawes and Constitutions of Scotland Faithfullie Collected.* London, 1609.

Somersetshire Pleas—from the Rolls of Itinerant Justices, ed. B. Healy. Somerset Records Society. London, 1897.

The Statutes of Ireland, Beginning the Third Yere of K. Edward the Second, and Continuing untill the End of the Parliament, Begunne in the Eleventh Yeare of the Reign of Our Most Gratious Soveraigne Lord King James. Dublin, 1621.

The Stonor Letters and Papers: 1290-1483, ed. Charles Lethbridge Kingsford. 2 vols. Camden Society Publication, Third Series, Vols. 29 and 30. London, 1919.

Stratton, Samuel W. *Report to the International High Commission Relative to the Use of the Metric System in Export Trade.* 64th Congress, 1st Session, Senate Document No. 241. Washington, D.C., 1916.

The Third Book of Remembrance of Southampton: 1514-1602, ed. A. L. Merson. Southampton, 1955.

Thomson, T., and Innes, C. *Acts of the Parliaments of Scotland, 1124-1707, with Supplement.* 11 vols. in 12. Edinburgh, 1875.

Thorpe, John. *Custumale Roffense, from the Original Manuscript in the Archives of the Dean and Chapter of Rochester.* London, 1788.

Three Rolls of the King's Court in the Reign of King Richard the First: A.D. 1194-1195. Pipe Roll Society. London, 1891.

Topham, John. *Wardrobe Account of the 28th Year of King Edward the First.* London, 1789.

"Weights, Measures and Coins." *Parliamentary Papers,* Great Britain: House of Commons Accounts and Papers, 58 (1864): 1-35.

Wellingborough Manorial Accounts: A.D. 1258-1323: from the Account Rolls of Crowland Abbey, ed. Frances M. Page. Northamptonshire, 1936.

The Welsh Port Books (1550-1603), ed. Edward Arthur Lewis. Cymmrodorion Record Series, Vol. 12. London, 1927. (Similar in format to the Southampton Port Books.)

The Whole Volume of Statutes at Large, Which at Anie Time Heeretofore Have Beene Extant in Print. London, 1587. (This is one of the earliest collections of the statutes. It checks out well against later editions and is notable primarily because of its unusual spellings.)

Wills and Inventories from the Registers of the Commissary of Bury St. Edmunds and the Archdeacon of Sudbury, ed. Samuel Tymms. Camden Society Publication, Vol. 49. London, 1850.

Wise, Charles. *The Compotus of the Manor of Kettering for A.D. 1292.* Kettering, 1899.

Year Books of Edward IV: 10 Edward IV and 49 Henry VI, ed. N. Neilson. Selden Society Publication, Vol. 47. London, 1931. (This volume contains occasional references to infractions of metrological regulations and to punishments imposed for violations of statutory provisions.)

York Memorandum Book: Part I (1376-1419). Surtees Society Publication, Vol. 120. Durham, 1912. (Thre are numerous references to weights and measures in addition to a table defining many linear and superficial units. Several documents are concerned with the proper sealing of weights and measures and with the scope of the mayor's jurisdiction in supervising regular assays.)

York Memorandum Book: Part II (1388-1493). Surtees Society Publication, Vol. 125. Durham, 1915. (The glossary has definitions for the selion and the butt of land.)

The York Mercers and Merchant Adventurers: 1356-1917. Surtees Society Publication, Vol. 129. Durham, 1918. (There are several documents that discuss the purchase of new weights and measures to replace the old and defective standards. The glossary contains information on the keel, fatt, and last.)

Reference Works

Akerman, John Yonge. *A Glossary of Provincial Words and Phrases in Use in Wiltshire.* London, 1842.

The Americana, ed. Frederick Converse Beach. Vol. 22. New York, 1912. (See article entitled "Weights and Measures.")

Ansileubus. "Glossarium." In *Glossaria Latina,* ed. W. M. Lindsay, J. F. Mountford, and J. Whatmough, 1:1-604. Paris, 1926. (There is an important entry for libra.)

Arts and Sciences or Fourth Division of the English Cyclopædia, ed. Charles Knight. Vol. 8. London, 1868. (See article entitled "Weights and Measures.")

Badcock, Benjamin. *Tables Exhibiting the Prices of Wheat, from the Year 1100 to 1830.* London, 1832.

Bailey, Nathan. *An Universal Etymological English Dictionary.* London, 1721.

Baker, Anne Elizabeth. *Glossary of Northamptonshire Words and Phrases.* 2 vols. London, 1854.

Bald, Alexander. *The Farmer and Corn-Dealers Assistant.* Edinburgh, 1780.

Baxter, J. H., and Johnson, Charles. *Medieval Latin Word-List from British and Irish Sources.* London, 1962.

Beck, S. William. *The Draper's Dictionary: A Manual of Textile Fabrics, Their History and Applications.* London, 1882.

Bradley, Richard, trans. *Chomel's Dictionaire aeconomique, or the Family Dictionary.* Translated and revised by R. Bradley. London, 1725.

Britten, James. *Old Country and Farming Words: Gleaned from Agricultural Books.* English Dialect Society, Vol. 30. London, 1880.

Brockett, John Trotter. *A Glossary of North Country Words in Use; with Their Etymology, and Affinity to Other Languages; and Occasional Notices of Local Customs and Popular Supersititions.* Newcastle upon Tyne, 1829. (Brockett includes entries for 15 weights and measures, and he discusses the characteristics of heaped, striked, and shallow capacity measures.)

Buckhurst, Helen McM. "An Anglo-Saxon Index." In *The Corpus Glossary,* ed. W. M. Lindsay, pp. 267-91. Cambridge, 1921. (Buckhurst does not define the weights and measures included in her list but only gives their declensions in Anglo-Saxon.)

Bullokar, John. *An English Expositor: Teaching the Interpretation of the Hardest Words Used in Our Language.* London, 1616. (Reproduced as No. 11 in the Collection of Facsimile Reprints of English Linguistics, 1500-1800, published by The Scolar Press Limited, London, 1967.)

Cawdrey, Robert. *A Table Alphabeticall, Conteyning and Teaching the True Writing, and Understanding of Hard Usuall English Wordes, Borrowed from the Hebrew, Greeke, Latine, or French.* London, 1604. (Reproduced, with an introduction by Robert A. Peters, and published by Scholars' Facsimiles & Reprints, Gainesville, Florida, 1966.)

Chambers, Ephraim. *Cyclopædia: or, An Universal Dictionary of Arts and Sciences.* Vol. 2. London, 1728.

──────. *Chambers's Encyclopædia.* Rev. ed., Vol. 10. London, 1874. (The sections on weights and measures in this edition are inferior to those in the earlier one.)

Chomel, M. Noel. *Dictionnaire œconomique.* Ed. P. Danjou. 4th rev. ed., 2 vols. Paris, 1740. (Includes separate entries for "mesures" and "poids" in Volume 2.)

Cocker, Edward. *Accomplish'd School-Master.* London, 1696. (Reproduced as No. 33 in the Collection of Facsimile Reprints of English Linguistics, 1500–1800, published by The Scolar Press Limited, London, 1967.)

Coles, E. *An English Dictionary, Explaining the Difficult Terms That Are Used in Divinity, Husbandry, Physick, Philosophy, Law, Navigation, Mathematicks, and Other Arts and Sciences.* London, 1724. (I also used the edition of 1732.)

Cooke, Layton. *Tables Adapted to the Use of Farmers and Graziers.* London, 1813. (A second edition was published in London in 1819 under the title *The Grazier's Manual.*)

Cooper, William Durrant. *A Glossary of Provincialisms in Use in the County of Sussex.* London, 1853. (Cooper's glossary contains some rather detailed accounts of unusual weights and measures such as draught, leap, meal, swod, tovet, warp, and wint.)

Courtney, W. S. *The Farmers' and Mechanics' Manual.* New York, 1880.

Cowell, John. *The Interpreter: or Booke Containing the Signification of Words.* Cambridge, 1607.

Cullyer, John. *The Gentleman & Farmer's Assistant.* London, 1798.

The Cyclopædia; or, Universal Dictionary of Arts, Sciences, and Literature, ed. Abraham Rees. Vol. 38. London, 1819. (See article entitled "Weight.")

Dickinson, William. *A Glossary of Words and Phrases Pertaining to the Dialect of Cumberland.* London, 1878.

Dictionarium rusticum, urbanicum et botanicum: or, A Dictionary of Husbandry, Gardening, Trade, Commerce, and All Sorts of Country-Affairs. London, 1717. (This dictionary contains approximately 100 entries for weights and measures. In several instances there are errors in computation and quite possibly a few printing errors.)

Dictionnaire universel français et latin. Paris, 1752.

Dinsdale, Frederick T. *A Glossary of Provincial Words Used in Teesdale in the County of Durham.* London, 1849. (Dinsdale has good definitions of noggin and score.)

Du Cange, Charles du Fresne. *Glossarium mediæ et infimæ Latinitatis.* 10 vols. Paris, 1937. (These volumes, a reissue of the original 1678 edition, contain a wealth of information on medieval English weights and measures. Not only are there definitions for some of them, but Du Cange includes ample documentation.)

The Edinburgh Encyclopædia, ed. David Brewster. Vols. 12 and 18. Philadelphia, 1832. (The first American edition; Volume 12 has an article on measures; Volume 18 on weights.)

Edler, Florence. *Glossary of Medieval Terms of Business: Italian Series 1200-1600.* Cambridge, 1934. (Although the weights and measures are Italian, there are references to their English equivalents in several instances.)

The Encyclopædia Britannica, or Dictionary of Arts, Sciences, and General Literature. 8th ed. Vol. 21. Edinburgh, 1860. (See article entitled "Weights and Measures.")

Encyclopédie méthodique. Vol. 3, Commerce. Paris, 1784. (There are many tables and charts comparing the metrological units of one country with another.)

Encyclopédie ou dictionnaire raisonné des sciences, des arts et des métiers, par une société de gens de lettres, ed. Denis Diderot. Vols. 21 and 26. Geneva, 1778. (Information on measures is in Volume 21 and on weights in Volume 26.)

Evans, John. *The Palace of Profitable Pleasure.* London, 1621. (Reproduced as No. 32 in the Collection of Facsimile Reprints of English Linguistics, 1500-1800, published by the Scolar Press Limited, London, 1967.)

Francis, Sidney. *Tables, Memoranda, and Calculated Results for Farmers, Graziers, Agricultural Students, Surveyors, Land Agents, Auctioneers, etc.* London, 1889. (I also used the 1890 and 1894 editions.)

"Glossarium Abba." In *Glossaria Latina,* ed. M. Inguanez and C. J. Fordyce, 5:9-143. Paris, 1931. (There are separate entries for stadium and dragma.)

Gouldman, Francis. *A Copious Dictionary.* London, 1664.

La Grande Encyclopédie: Inventaire raisonné des sciences, des lettres et des arts, ed. MM. Berthelot and Laurent. Paris, 1886-1902. Vol. 26. (This volume contains one small section on the weights and measures of the Middle Ages.)

Gregory, G. *A Dictionary of Arts and Sciences.* 3 vols. New York, 1822. (Volume 2 has an article on measures, volume 3 on weights.)

Harris, John. *Lexicon technicum, or, An Universal English Dictionary of Arts and Sciences.* Vol. 2. London, 1710. (Information is included under the entries "weights" and "measures.")

———. *Lexicon technicum, or, An Universal English Dictionary of Arts and Sciences.* (3d ed.) London, 1716. (This work is much more detailed than the earlier edition and contains some excellent tables comparing English weights and measures with ancient and contemporary systems.)

Hatch, F. H., and Vallentine, E. J. *Mining Tables: Being a Comparison of the Units of Weight, Measure, Currency, Mining Area, etc., of Different Countries; Together with Tables, Constants & Other Data Useful to Mining Engineers and Surveyors.* London, 1907.

Hilderbrand, Clifton. *Metric Literature Clues: A List of References to Books, Pamphlets, Documents and Magazine Articles on Metric Standardization of Weights and Measures.* San Francisco, 1921. (This collection is somewhat dated.)

Hofmann, Joh. Jacob. *Lexicon universale historiam sacram et profanam.* Leiden, 1698.

Holme, Randle. *The Academy of Armory, or a Storehouse of Armory and Blazon.* London, 1688.

Hunter, Rev. Joseph. *The Hallamshire Glossary.* London, 1824.

Huntley, Rev. Richard Webster. *A Glossary of the Cotswold (Gloucestershire) Dialect.* London, 1868. (Huntley discusses the lug.)

International Traders' Handbook. Philadelphia, 1934.

Johnson, Samuel. *A Dictionary of the English Language.* 2 vols. London, 1773. (Although there are many entries for weights and measures, there are occasional errors in both their size and composition.)

Johnson's Universal Cyclopædia, ed. Charles Kendall Adams. Vol. 8. New York, 1895. (See article entitled "Weights and Measures.")

Labbe, Philippe. *Bibliotheca bibliothecarum curis secundis auctior.* Rouen, 1672.

Lambart, James. *The Countrymans Treasure.* London, 1683.

Leigh, Egerton. *A Glossary of Words Used in the Dialect of Cheshire.* London, 1877. (Leigh includes separate entries for 9 weights and measures.)

McConnell, Primrose. *Note-Book of Agricultural Facts & Figures for Farmers and Farm Students.* London, 1883.

Morton, John C. *A Cyclopedia of Agriculture Practical and Scientific.* London, 1855.

Murphy, Edmund. *The Agricultural Instructor or Farmer's Class-Book.* Dublin, 1849.

Owen, John. *The Youth's Instructor in the English Tongue.* London, 1732. (Reproduced as No. 14 in the Collection of Facsimile Reprints of English Linguistics, 1500–1800, published by The Scolar Press Limited, London, 1967.)

The People's Cyclopedia of Universal Knowledge, with Numerous Appendixes Invaluable for Reference in All Departments of Industrial Life, ed. William Harrison De Puy. Vol. 3. New York, 1885.

Phillips, Edward. *The New World of English Words; or, A General Dictionary.* London, 1696. (I also used the edition of 1706.)

Pigott, I. *The Canadian Mechanic's Ready Reckoner.* Three Rivers, Canada, 1832.

Pope, Charles. *The Merchant, Ship-Owner, and Ship-Master's Import and Export Guide.* London, 1831.

Postlethwayt, Malachy. *The Universal Dictionary of Trade and Commerce.* 2 vols. London, 1755. (Weights and measures information is in Volume 2, pp. 186–97.)

Promptorium Parvulorum, ed. Albertus Way. Camden Society Publication. London, 1843.

Rastell, John. *An Exposition of Certaine Difficult and Obscure Words and Termes of the Lawes of This Realme.* London, 1579.
Ray, John. *A Collection of English Words Not Generally Used.* London, 1674.
Renton, George. *The Graziers' Ready Reckoner.* London, 1804.
Rider, John. *Riders Dictionarie, Corrected and Augmented with the Addition of Many Hundred Words Both Out of the Law, and Out of the Latine, French, and Other Languages.* London, 1640.
Robinson, Francis Kildale. *A Glossary of Yorkshire Words and Phrases Collected in Whitby and the Neighbourhood.* London, 1855.
Rolt, Richard. *A New Dictionary of Trade and Commerce.* London, 1756.
Sandys, Sir John Edwin. *A Companion to Latin Studies.* 3d ed. Cambridge, 1921. (Sandys discusses Roman metrology rather briefly.)
Sheppard, James. *The British Corn Merchant's and Farmer's Manual.* Derby, 1820.
Shipley, Joseph T. *Dictionary of Early English.* New York, 1955. (Shipley defines some unusual units such as the fust, fardel, and seron.)
Simmonds, P. L. *The Commercial Dictionary of Trade Products, Manufacturing and Technical Terms: with a Definition of the Moneys, Weights, and Measures of All Countries Reduced to the British Standard.* London, 1883.
Skene, Sir John. *De verborum significatione.* London, 1609.
Skilling, Thomas. *The Farmer's Ready Reckoner or Glasnevin Agricultural Tables.* Dublin, 1848.
Spelman, Henry. *Glossarium archaiologicum.* London, 1664.
Sternberg, Thomas. *The Dialect and Folk-Lore of Northamptonshire.* London, 1851. (Sternberg has entries for 8 measures.)
Tap, John. *The Seamans Kalender, or, An Ephemerides of the Sunne, Moone, and Certaine of the Most Notable Fixed Starres.* 9th ed. London, 1625.
Waterston, William. *A Manual of Commerce.* Edinburgh, 1840.
Wedgwood, Hensleigh. *A Dictionary of English Etymology.* London, 1878. (Wedgwood gives etymological derivations for 53 units.)
Wilkinson, Robert Oliver. *The Druggist's Price-Book, or a Catalogue of the Drugs, Chemicals, & Perfumery, Generally Sold by Chemists & Druggists, with the Doses and Old Names Annexed.* 2d ed. London, 1832.

Other Works

Adames, Jonas. *The Order of Keeping a Court Leete and Court Baron.* London, 1593.
Andrews, Charles McLean. *The Old English Manor.* Baltimore, 1892.
Astle, Thomas. *An Account of the Tenures, Customs, &c. of the Manor of Great Tey, in the County of Essex.* London, 1795.
Ault, Warren O. "Open-Field Husbandry and the Village Community." *Transactions of the American Philosophical Society,* 55 (1965): 1–102. (The appendix contains by-laws taken from the rolls of 37 manors in 12 counties, dating from 1270 to 1608.)

B. J. *The Merchants Avizo*. London, 1607. (The author was a merchant.)
Baker, Humfrey. *The Well-Spring of Sciences*. London, 1646. (Tables of weights and measures.)
Barbon, Nicholas. *A Discourse of Trade*. London, 1690.
Barfield, Samuel. *Thatcham, Berks, and Its Manors*. Oxford and London, 1901.
Barrington, Daines. *Observations on the More Ancient Statutes from Magna Carta to the Twenty-First James I*. London, 1796.
Beard, Charles Austin. *The Office of Justice of the Peace in England: In Its Origin and Development*. New York, 1904.
Beawes, Wyndham. *Lex Mercatoria Rediviva; or, The Merchant's Directory*. London, 1783. (Discussions and tables of weights and measures.)
Beck, S. William. *Gloves, Their Annals and Associations: A Chapter of Trade and Social History*. London, 1883. (Beck includes many documents illustrating the various metrological units found in glove manufacture and trade.)
Bellot, James. "The Booke of Thrift, Containing a Perfite Order, and Right Methode to Profite Lands, and Other Things Belonging to Husbandry." In *The Manor Farm*, ed. Francis Henry Cripps-Day, pp. 115 ff. London, 1931. (Reprint of original 1589 edition.)
Boissonnade, P. *Life and Work in Medieval Europe*. New York, 1950. (Boissonnade mentions the hogshead and the pound.)
Bolton, Richard. *A Iustice of Peace for Ireland, Consisting of Two Bookes*. Dublin, 1638.
Bonwick, James. *Romance of the Wool Trade*. London, 1887.
Bourquelot, M. Felix. *Etudes sur les Foires de Champagne*. Paris, 1865. (Bourquelot devotes one entire chapter to the weights and measures used by merchants at the fairs of Champagne.)
Bréhaut, Ernest. *An Encyclopedist of the Dark Ages: Isidore of Seville*. New York, 1912. (Bréhaut includes a translation of Isidore's short treatise on weights and measures.)
Bridbury, A. R. *England and the Salt Trade in the Later Middle Ages*. Oxford, 1955.
Bridges, Noah. *Lux Mercatoria: Arithmetick Natural and Decimal*. London, 1660.
Brown, Cornelius. *A History of Newark on Trent*. Newark, 1904.
Budé, Guillaume. *Annotationes Gulielmi Budæ Parisiensis, secretarii regii, in quatuor et viginti pandectarum libros, ad Io annem deganaium cancellarium Franciæ*. Paris, 1535. (The weights and measures discussed are Greek and Roman.)
Carew, Richard. *The Survey of Cornwall*. London, 1602.
──────. *Carew's Survey of Cornwall; to Which Are Added, Notes Illustrative of Its History and Antiquities*. Ed. Thomas Tonkin and Francis Lord de Dustanville. London, 1811.
Celsus, Aulus Corn. *Medicina libri octo*. Leipzig, 1766.

Other Works

Chadwick, Hector Munro. *Studies on Anglo-Saxon Institutions.* New York, 1963. (Chadwick discusses the pound. The book is a reissue of the original 1905 edition.)

Chamberlayne, John. *Magna Britannia Notitia; or, The Present State of Great-Britain with Divers Remarks upon the Ancient State Thereof.* London, 1708. (One chapter is devoted to weights and measures. Chamberlayne is rather repetitious and sometimes quotes materials without indicating his sources. He provides, however, many useful tables illustrating the dimensions of superficial and capacity measures.)

Child, Joseph. *A Discourse concerning Trade.* London, 1689.

Clark, George Thomas. "The Custumary of the Manor and Soke of Rothley, in the County of Leicester." *Archaeologia,* 47 (1882): 89–130.

Clode, Charles M. *The Early History of the Guild of Merchant Taylors of the Fraternity of St. John the Baptist, London, with Notices of the Lives of Some of Its Eminent Members: Part 1 (The History).* London, 1888. (Some valuable documents.)

Coke, Edward. *The Fourth Part of the Institutes of the Laws of England concerning the Jurisdiction of the Courts.* London, 1797.

"Common Rights at Cottenham and Stretham in Cambridgeshire." In *Camden Miscellany,* ed. W. Cunningham. *Camden Third Series,* 12 (1910): 169–290.

Cortés, Martin. *The Arte of Nauigation.* Tr. Richard Eden. London, 1561.

Coulton, G. G. *Medieval Village, Manor and Monastery.* New York, 1960. (Coulton discusses rather briefly the perch, sheaf, thrave, and yardland and lists some of the problems resulting from variations in these measures.)

Cunningham, William. *The Growth of English Industry and Commerce during the Early and Middle Ages.* Cambridge, 1927. (Cunningham defines briefly 12 weights and measures.)

Curtler, W. H. R. *A Short History of English Agriculture.* Oxford, 1909.

Dalton, Michael. *The Countrey Justice.* London, 1635. (One entire chapter is devoted to weights and measures. Especially valuable is the discussion of the duties and responsibilities of the justices of the peace in regard to verification and enforcement of Crown standards.)

Davenport, Frances Gardiner. *The Economic Development of a Norfolk Manor: 1086–1565.* Cambridge, 1906. (Especially valuable are the appendixes of documents.)

Dickinson, William Croft. *Scotland from the Earliest Times to 1603.* Edinburgh, 1962.

Digges, Dudley. *The Defense of Trade: In a Letter to Sir Thomas Smith Knight, Gouernour of the East-India Companie.* London, 1615.

Digges, Leonard. *A Booke Named Tectonicon.* London, 1647. (Digges provides very little information on individual units aside from defining the inch, yard, and perch when setting up specific arithmetical problems for the reader to solve.)

Dugdale, Sir William. "Monasticon Anglicanum." In *The Old Historians of the Isle of Man,* ed. William Harrison Douglas, pp. 1-77. Isle of Man, 1871. (There is some treatment of superficial measures in the entry for Rushen Abbey, p. 75.)

Eliot, F. Perceval. *Letters on the Political and Financial Situation of the Country in the Year 1814; Addressed to the Earl of Liverpool.* London. 1814. (Author uses the pseudonym Falkland.)

Elton, Charles Isaac. *The Tenures of Kent.* London, 1867. (Chapter 6, "The Domesday Survey," is valuable for land measures.)

Ewart, John. *The Land Drainer's Calculator.* London, 1862.

Eyre, J. *The Exact Surveyor; or, The Whole Art of Surveying of Land.* London, 1654.

Eyton, Rev. Robert William. *Domesday Studies: An Analysis and Digest of the Somerset Survey (according to the Exon Codex), and of the Somerset Gheld Inquest of A.D. 1084.* 2 vols. London, 1880. (Eyton discusses the hide as a unit of superficial measurement.)

Fitzherbert, Anthony. *The Boke of Hvsbandry.* London, 1534.

_____. *The Book of Husbandry by Master Fitzherbert.* Ed. Rev. Walter W. Skeat. London, 1882. (An edited version of the original. Issued by the English Dialect Society.)

Fleetwood, Bishop. *Chronicon preciosum; or, An Account of English Gold and Silver Money; the Price of Corn and Other Commodities; and of Stipends, Salaries, Wages, Jointures, Portions, Day-Labour, etc. in England, for Six Hundred Years Last Past.* London, 1745. (Fleetwood discusses 14 separate units of measurement, but his treatment is very superficial and he confuses the troy with the tower pound. He does, however, have some timely quotations from medieval and early modern manuscripts.)

Folkingham, W. *Fevdigraphia: The Synopsis or Epitome of Svrveying Methodized.* London, 1610.

Forbes, William. *The Duty and Powers of Justices of Peace, in This Part of Great Britain Called Scotland; with an Appendix concerning Weights and Measures.* Edinburgh, 1707.

Fox, Francis F. *Some Account of the Ancient Fraternity of Merchant Taylors of Bristol.* London, 1880.

Gras, Norman Scott Brien. *The Early English Customs Systems.* Cambridge, 1918. (Occasionally a certain capacity measure is defined, but generally only the price for its contents is given.)

_____. *The Economic and Social History of an English Village (Crawley, Hampshire) A.D. 909-1928.* Cambridge, 1930.

_____. *The Evolution of the English Corn Market from the Twelfth to the Eighteenth Century.* Cambridge, 1915. (Contains a short description of the seam.)

Gray, Dionis. *The Store-House of Breuitie in Woorkes of Arithemetike.* London, 1577.

Gray, Howard Levi. *English Field Systems.* Cambridge, 1915. (The appendixes contain some excellent documents.)

Greenstreet, James. *Assessments in Kent for the Aid to Knight the Black Prince: Anno 20 Edward III.* London, 1878.

Greenwood, William. *The Authority, Jurisdiction and Method of Keeping County-Courts, Courts-Leet, and Courts-Baron.* London, 1730.

Gross, Charles. *The Gild Merchant.* 2 vols. Oxford, 1890. (A rich source for documentation.)

Hale, Lord Chief Justice. "A Treatise, in Three Parts. Pars Prima. De Jure Maris et Brachiorum ejusdem. Pars Secunda. De Portibus Maris. Pars Tertia. Concerning the Customs of Goods Imported and Exported." In *A Collection of Tracts Relative to the Law of England, from Manuscripts, Now First Edited,* ed. Francis Hargrave, Vol. 1. Dublin, 1787. (Especially important is Part 3, pp. 115-248.)

Harpur, John. *The Iewell of Arithmetick.* London, 1617.

Hatton, Edward. *Arithmetick; or, The Ground of Arts: Teaching that Science, Both in Whole Numbers and Fractions.* London, 1699. (A revised version of the work originally written by Robert Recorde. See below, under Recorde.)

An Intire System of Arithmetic; or, Arithmetic in All Its Parts. London, 1731.

―――. *The Merchant's Magazine; or, Trades-Man's Treasury.* London, 1701.

Hazlitt, W. Carew, ed. *Tenures of Land & Customs of Manors.* London, 1874. (Originally written by Thomas Blount; rearranged, corrected, and enlarged from the copies of Blount [1679], Beckwith [1784], and the third edition of 1815.)

Henderson, David. *Tables for Calculating the Price of any Quantity of Grain.* Edinburgh, 1838.

Hewitt, H. J. *Medieval Cheshire: An Economic and Social History of Cheshire in the Reigns of the Three Edwards.* Chetham Society Publication, Vol. 88. Manchester, 1929. (Appendix G describes some of the linear, superficial, and capacity measures that were found in Cheshire.)

Hill, Thomas. *The Art of Vulgar Arithmeticke, Both in Integers and Fractions, Deuided into Two Bookes.* London, 1600.

Hodder, James. *Hodder's Arithmetick; or, That Necessary Art Made Most Easie.* London, 1661.

Hopton, Arthur. *A Concordancy of Yeares.* London, 1616. (Chapter 43 deals with weights and measures.)

Howes, Edward. *Short Arithmetick; or, The Old and Tedious Way of Numbering, Reduced to a New and Briefe Method.* London, 1656.

Hunt, Nicholas. *The Merchants Iewell.* London, 1628.

Hylles, Thomas. *The Arte of Vulgar Arithmeticke.* London, 1600.

Introduction to the Curia Regis Rolls, 1199-1230. A.D., ed. Cyril Thomas Flower. Selden Society Publication, Vol. 62. London, 1944. (This work contains several references to land measures.)

Jager, Robert. *Artificial Arithmetick in Decimals.* London, 1651.
Jeake, S. *Arithmetic.* London, 1696.
Jenkins, David. *Pacis Consultum: A Directory to the Publik Peace.* London, 1657.
Jolliffe, J. E. A. *Constitutional History of Medieval England.* London, 1954.
Kennett, White. *Parochial Antiquities Attempted in the History of Ambrosden, Burcester, and Other Adjacent Parts in the Counties of Oxford and Bucks.* Oxford, 1695. (The book has a glossary containing some definitions of the weights and measures found in these counties.)
Kern, F. *Kingship and Law in the Middle Ages.* Oxford, 1956.
King, Charles. *The British Merchant; or, Commerce Preserv'd.* 2 vols. London, 1721. (Volume 1 is important for information on weights and measures employed in the European trade.)
Kytchin, John. *Le Covrt Leete et Court Baron.* London, 1580.
Lavoisier, Antoine Laurent. *Œuvres.* Ed. René Fric. 2 vols. Paris, 1955. (Lavoisier was one of the pioneers in the construction of the metric system, which put an end to the complexity and confusion of French weights and measures. These two volumes contain some of his writings on early modern measuring units.)
_____. *Statistique agricole et projets de réformes.* Ed. Edouard Grimaux. Paris, 1888.
Leigh, Valentine. *Moste Profitable and Commendable Science of Surueying of Landes, Tenementes, and Hereditamentes.* London, 1577.
Levi, Leone. *The History of British Commerce and of the Economic Progress of the British Nation: 1763-1878.* London, 1880.
Leybourn, William. *The Compleat Surveyor.* London, 1653.
_____. *Planometria; or, The Whole Art of Svrveying of Land.* London, 1650.
Lightwood, John M. *A Treatise on Possession of Land.* London, 1894.
Lipson, E. *The Economic History of England.* 2 vols. London, 1949. (In volume 1 Lipson discusses the supervision of weights and measures by state and local officials.)
Lloyd, J. E. *Early Welsh Agriculture.* Bangor, 1894.
Loch, David. *Essays on the Trade, Commerce, Manufactures, and Fisheries of Scotland.* 3 vols. Edinburgh, 1778-1779.
Lyte, Henry. *The Art of Tens, or Decimall Arithmeticke.* London, 1619.
Macpherson, David. *Annals of Commerce, Manufactures, Fisheries, and Navigation with Brief Notices of the Arts and Sciences Connected with Them.* 4 vols. London, 1805. (Volume 1 has some excellent tables of medieval English measuring units.)
Maitland, Frederic William. *Domesday Book and Beyond: Three Essays in the Early History of England.* New York, 1966. (First published in 1897 in Cambridge, England, and Boston, Massachusetts.)
Mariana, Juan de. *De ponderibus et mensuris.* N.p., 1611.
Marshall, William H. *The Rural Economy of Yorkshire.* London, 1788.

Mason, R. *A Mirrorr for Merchants.* London, 1609.
Masterson, Thomas. *Thomas Masterson His First (-Third) Booke of Arithmeticke.* London, 1592.
Mayne, John. *Arithmetick: Vulgar, Decimal, & Algebraical.* London, 1675.
Montgomery, William Ernest. *The History of Land Tenure in Ireland.* Cambridge, 1889.
Murray, K. M. *A Constitutional History of the Cinque Ports.* Manchester, 1935.
Neilson, Nellie. *Economic Conditions on the Manors of Ramsey Abbey.* Philadelphia, 1898. (Valuable documents in the appendix.)
Nicolson, Joseph, and Burn, Richard. *The History and Antiquities of the Counties of Westmorland and Cumberland.* 2 vols. London, 1777. (The glossary in volume 2 has definitions for several superficial measures.)
Norden, John. *The Surveiors Dialogue.* London, 1610.
Nourse, T. *Campania Fœlix; or, A Discourse of the Benefits and Improvements of Husbandry.* London, 1700.
Noy, R. *Complete Lawyer.* London, 1634. (A 1651 edition is entitled *The Compleat Lawyer, or a Treatise Concerning Tenvres and Estates.*)
Owen of Henllys, George. *The Description of Penbrokshire.* Ed. Henry Owen. 3 vols. London, 1892. (Owen's book is valuable for Welsh linear, superficial, and capacity measures, and his description of the perch is especially important.)
Palladius. *On Husbondrie.* Ed. Rev. Barton Lodge. Early English Text Society. London, 1873.
Paton, Joseph N. "Linen Manufactures." *Encyclopædia Britannica.* 9th ed., 1888. 14: 633-68.
Pearman, M. T. *A History of the Manor of Bensington.* London, 1896.
Pegolotti, Francesco Balducci. *La Pratica della mercatura,* ed. Allan Evans. Cambridge, 1936. (Pegolotti describes the clove, hundred, and stone in addition to many non-English units.)
Pell, O. C. "Summary of a New View of the Geldable Unit of Assessment of Domesday." In *Domesday Studies,* ed. P. Edward Dove, 2:561-619. London, 1888.
Powell, John. *The Assize of Bread.* London, 1595.
Powell, Robert. *A Treatise of the Antiquity, Authority, Vses and Jurisdiction of the Ancient Courts of Leet.* London. 1642.
Rathborne, Aaron. *The Svrveyor in Foure Bookes.* London, 1616.
Rawlyns, Richard. *Practical Arithmetick.* London, 1656.
Recorde, Robert. *The Ground of Artes, Teachyng the Worke and Practice of Arithmetike.* London, 1540. (I also used the edition of 1566.)
Ricard, Samuel. *Traité général du Commerce.* 2 vols. Amsterdam, 1781. (Volume 2, pp. 150-56, discusses English weights and measures.)
Roberts, Lewes. *The Merchants Map of Commerce.* London, 1677.
Robertson, E. William. *Historical Essays in Connexion with the Land, the Church, &c.* Edinburgh, 1872.

Robinson, Thomas. *The Common Law of Kent: The Customs of Gavelkind.* London, 1897.
Rogers, James E. Thorold. *A History of Agriculture and Prices in England.* Vol. 1. Oxford, 1886. (Rogers describes 34 weights and measures and examines the standards upon which they are based. His descriptions, however, are not detailed.)
———. *Six Centuries of Work and Wages.* New York, 1884. (Roger discusses Arabic numbers, the hide, and the gallon.)
Rördansz, C. W. *European Commerce: or, Complete Mercantile Guide to the Continent of Europe.* London, 1818.
Round, J. H. "Barons and Knights in the Great Charter." In *Magna Carta Commemoration Essays.* London, 1917.
———. *Feudal England.* London, 1895. (The first part of the book consists of studies of the Domesday Book and of similar but less important surveys; the second part, of essays embodying the results of the author's researches in the history of this period.)
Salignacus, Bernardus. *The Principles of Arithmeticke.* Tr. W. Bedwell. London, 1616.
Salzman, L. F. *English Industries of the Middle Ages.* Oxford, 1923. (Salzman discusses the various capacity measures that were used for coal, iron, fish, and malted beverages.)
———. *English Trade in the Middle Ages.* Oxford, 1931. (One chapter is devoted to English weights and measures. Most of Salzman's remarks are really too brief to be sufficiently informative, and his most important contribution is his section on Roman and Arabic methods of computation.)
Sandys, Charles. *Consuetudines Kanciae: A History of Gavelkind and Other Remarkable Customs in the County of Kent.* London, 1851.
Scrope, G. Poulett. *History of the Manor and Ancient Barony of Castle Combe in the County of Wilts.* London, 1852. (Many important rental rolls and court records.)
A Select Collection of Early English Tracts on Commerce from the Originals of Mun, Roberts, North and Others, ed. John Ramsay McCulloch. Cambridge, 1952.
Slade, C. F. *The Leicestershire Survey: c. A.D. 1130.* Preface by Frank M. Stenton. No. 7 in the Department of English Local History Occasional Papers, ed. H. P. R. Finberg. Leicester, 1956.
Stevens, A. B. *Arithmetic of Pharmacy.* 4th ed. New York, 1920.
Stevin, Simon. *Disme: The Art of Tenths, or Decimall Arithmetike.* Tr. R. Norton. London, 1608.
Stubbs, William. *Constitutional History.* Oxford, 1906.
———. *Historical Introduction to the Rolls Series.* London, 1902.
Tait, James. *The Medieval English Borough.* Manchester, 1836.
Tap, John. *The Path-Way to Knowledge; Containing the Whole Art of Arithmeticke, Both in Numbers and Fractions.* London, 1613.

Taylor, Isaac. "The Ploughland and the Plough." In *Domesday Studies,* ed. P. Edward Dove, 1:143-88. London, 1888.

Turgot, Anne Robert Jacques. *Œuvres de Turgot.* Ed. Eugène Daire. 2 vols. Paris, 1844.

Tusser, Thomas. *Fiue Hundreth Pointes of Good Husbandrie.* English Dialect Society. London, 1878.

Violet, Thomas. *The Advancement of Merchandize.* London, 1651.

Walter of Henley's Husbandry: Together with an Anonymous Husbandry, Seneschaucie and Robert Grosseteste's Rules, ed. Elizabeth Lamond. London, 1890. (There is information on the furlong, perch, and league in Walter's work; in the *Anonymous Husbandry* there is a discussion of the perch, acre, and rood.)

Warburton, Rev. W. *Edward III.* London, 1924. (Contains short definitions of the sack and pack of wool.)

Weinbaum, Martin. "London unter Eduard I und II." *Vierteljahrschrift für Sozial und Wirtschaftsgeschichte,* Supplement 29 (1933). (There is information on the Assize of Bread and the clove.)

Wheeler, John. *A Treatise of Commerce, Wherin Are Shewed the Commodities Arising By a Wel Ordered, and Rvled Trade.* Middelburgh, 1601. (Republished in 1931 by the New York University Press, as edited by George Burton Hotchkiss.)

Wigan, Eleazar. *Practical Arithmetick.* London, 1695.

"William Gregory's Chronicle of London." In *The Historical Collections of a Citizen of London in the Fifteenth Century,* ed. James Gairdner, pp. 55-239. Camden Society Publication. London, 1876.

Willis, Browne. *The History and Antiquities of the Town, Hundred, and Deanry of Buckingham.* London, 1755. (Willis discusses the larger superficial measures such as the virgate, bovate, plowland, and hide.)

Winter, George. *A Compendious System of Husbandry.* London, 1797.

Wood, William. *A Survey of Trade.* London, 1718.

Worlidge, John. *Systema agriculturæ; the Mystery of Husbandry Discovered.* London, 1669.

Young, William. *The History of Dulwich College.* 2 vols. Edinburgh, 1889. (Volume 2 contains some interesting variant spellings of weights and measures in a list of pharmaceutical recipes and in the diary of Edward Allen.)

Addendum

Three recent publications in British metrology should be of special value to those interested in the proposed changeover to the metric system. The page numbers in parentheses after each entry refer to additional bibliographical information contained in these books.

Anderton, Pamela, and Bigg, P. H. *Changing to the Metric System: Conversion Factors, Symbols and Definitions.* London, 1969. (pp. 32-33)

Barrell, H. "A Short History of Measurement Standards at the National Physical Laboratory." *Contemporary Physics,* 9 (1968): 205–26. (pp. 225–26)

National Physical Laboratory. *NPL Historical Exhibition of Standards and Measuring Equipment.* London, 1967. (pp. 4, 5, 6, 7, 9, 10, 12 and 14)

In the United States, the following books, articles, and pamphlets published since 1956 are especially important for their tie-in to British metrology, for their information on American efforts at solving some of the problems of the British system, and for their detailed expositions defining the difficulties faced in preparatory adoption of the metric system. Again, page numbers in parentheses after some of the entries will supply additional bibliographical aids. The abbreviations NBS represents the National Bureau of Standards.

Bussey, W. S., and Jensen, M. W. *The Development of Weights and Measures Control of Packaged Goods in the United States.* Washington, D.C., 1960. (p. 7)

———. *Weights and Measures Administration in the United States.* Washington, D.C., 1956.

Chisholm, L. J., ed. *Report of the 51st National Conference on Weights and Measures 1966.* NBS Miscellaneous Publication No. 290. Washington, D.C., 1967. (Bibliography on back cover.)

———. *Units of Weight and Measure: International (Metric) and U.S. Customary.* NBS Miscellaneous Publication No. 286. Washington, D.C., 1967. (p. 8)

Hughes, J. C., and Keysar, B. C. *Testing of Metal Volumetric Standards.* NBS Monograph No. 62. Washington, D.C., 1963.

Judson, Lewis V. *Weights and Measures Standards of the United States: A Brief History.* NBS Miscellaneous Publication No. 247. Washington, D.C., 1963. (p. 25)

McPherson, Archibald T. *Problems Involved in Determining the Cost and the Optimum Time of Conversion to the Metric System.* Washington, D.C., 1962.

NBS. *ASTM Metric Practice Guide. NBS Handbook No. 102.* Washington, D.C., 1967 (pp. 44–45)

———. *Brief History and Use of the English and Metric Systems of Measurement.* NBS Special Publication No. 304A. Washington, D.C., 1968.

———. *List of State, Commonwealth, and District Weights and Measures Offices of the United States.* Washington, D.C., 1968.

———. *Model Regulation Pertaining to the Voluntary Registration of Servicemen and Service Agencies for Commercial Weighing and Measuring Devices.* Washington, D.C., 1966.

———. *Model State Law on Weights and Measures: Form 2.* Washington, D.C., 1968.

———. *Model State Packaging and Labeling Regulation of 1968 as Adopted by the National Conference on Weights and Measures.* Washington, D.C., 1968.

_____. *Model State Weighmaster Law.* Washington, D.C., 1965.

_____. *The National Conference on Weights and Measures.* Washington, D.C., 1965.

_____. *Present Situation regarding the Use of the Metric System in the United States.* Washington, D.C., 1968.

_____. *Weights and Measures Administration.* NBS Handbook No. 82. Washington, D.C., 1962.

Smith, Ralph W. *The Federal Basis for Weights and Measures.* NBS Circular No. 593. Washington, D.C., 1958. (p. 23)

Zupko, Ronald Edward. "The Long Ordeal: Conversion to the Metric System in the United States." *Intellect,* 105 (September-October 1976): 114–17.

_____. "The Metric Conversion: Transition from the Pre-Imperial to the Metric System in the United States." *Marquette Business Review,* 17 (1973): 49–58.

_____. "The Origin and Development of the Metric System in France: An Historical Outline." *Metric System Guide Bulletin,* 11 (1974): 25–31.

_____. "A Short History of the Metric System in the United States: 1790 to 1974." *Metric System Guide Bulletin,* 12 (1974): 3–9.

_____. "Worldwide Dissemination of the Metric System During the Nineteenth and Twentieth Centuries." *Metric System Guide Bulletin,* 9 (1974): 14–25.

Index

Abbesses, 49
Abbots, 35, 48–49
Aberdeen, 30
Acre: of North German system, 11; in Composition of Yards and Perches, 21
Actus, 7
Actus quadratus, 7
Adams, John Quincy, 15n
Aelle, Bretwalda, 10
Agricola, Julius, 3
Agriculture, cash-crop, 71
Airy Commission, 96
Alaric, King of Visigoths, 4
Aldermen: metrological duties of, 42; barred clerks of the market from cities, 69; coopers appeal to, 84; orders for burning of defective casks, 84; mentioned, 44, 47, 50, 82. See also Urban officials
Alfred, king of Wessex, 10, 12
Alnage, 59, 82
Alnagers: in statute of 1405, 28; in statute of 1328, 44, 60; functions and metrological duties of, 59–62; in statute of 1351, 60; later statutes dealing with, 61–62; height and decline of power of, 62; sources on, 62; infractions committed by, 62n
Amphora, 8
Ampleford, John de, 66–67
Angevins, 71
Anglo-Saxon Britain. See Britain, Anglo-Saxon
Anne, queen of England, 95
Antonine Wall, 4
Appleby, 75n
Appleby, Nicholas de, 41
Apulia, 15
Archbishops, 35, 49
As, 7
Assays: of coins, 15n; distinguished from assizes, 17n; of clerks of the market, 68. See also Assizes
Assisa mensurarum. See Assize of Measures
Assisa panis et cervisiae. See Assize of Bread and Ale
Assize of Bread and Ale (*Assisa panis et cervisiae*): description of, 19–20; political and commercial implications of, 20; urban residents' metrological functions in, 36, 38; problems with officials' metrological duties in, 36–37; mentioned, 20
Assize of Measures (*Assisa mensurarum*): description of, 18; importance of, 36; urban residents' metrological functions in, 36; problems with, 36
Assizes: characteristics of, 17; definition of, 17n; metrological functions contained in, 48, 53. See also Assays
Auncel: in Statute of Purveyors, 23; lifting of, 24n; abolished in 1360, 25; reasons for continued use, 25n, 26n, 30; third condemnation of, 30
Aune, 61. See also Ell
Avignon, 71
Avoirdupois, 25. See also Ounce; Pound; Weights

Babylonian Captivity, 71
Bailiffs: presentments before, 35; commissioners work with, 37; election of, 41n; metrological duties of, 42, 43; of manorial courts, 46; assistants to coroners, 53; assistants to sheriffs, 53; warrants to, 70; mentioned, 41, 47, 60, 82. See also Urban officials
Bala, 89
Balance: in Statute of the Staple, 24; description of, 24n, 25n; purchase

235

Balance (*cont.*)
  of, 25n; in statute of 1429, 30; in statute of 1357, 32; mentioned, 32. *See also* Beam; Scales
Banbury, Nicholas de, 40
Barge-load, 29
Barleycorns: as division of North German thumb, 10; in Composition of Yards and Perches, 21; in sterling or pennyweight, 77. *See also* Inch
Barons, 14
Barrel: in statute of 1423, 29; of herrings, 58; of eels, 58; variations of, 82; of ale and beer, 82; of soap, 82–83
Baynard, Robert, 40
Beadles, 53
Beam: maintenance of, 64; "great," 25n; "king's," 25n; "small," 25n. *See also* Balance; Scales
Bedford, 75n
Bekyngton, John de, 41
Berkshire, 54n
Berriman, A. E., 7n
Beverly, 34
Bina jugera, 7
Birmingham, 73
Biroun, Edmund, 65
Bishops, 49
Black Death, 71
Black Prince, 47
Blast furnace, 72, 73
Board of Trade: Standards Department of, 80; efforts of, in metric changeover, 98
Boston, 89
Brecon, borough of, 89
Bremen pound, 29
Bretwalda system, 11
Brewers, 82
Brigantes, 3
Bristol, 43, 75n
Britain
  —Anglo-Saxon: characteristics of, 9–10; linear measures of, 10–11; weights and measures units of, 10–14 *passim;* weights and measures standards of, 10, 12–14 *passim;* superficial measures of, 11; weights of, 11; capacity measures of, 14; weights and measures bequests of,

16. *See also* Standards
  —Celtic: weights and measures units of, 5–6; weights and measures standards of, 6
  —Norman: weights and measures units and standards of, 14; retention of Saxon metrological system in, 14; retention of Saxon institutions in, 15; weights and measures characteristics of, 16. *See also* William I, king of England
  —Roman: characteristics of, 3–5; weights and measures units of, 5–9 *passim;* weights and measures standards of, 5, 8; linear measures of, 6; itinerary measures of, 6–7; superficial measures of, 7; weights of, 7–8; capacity measures of, 8; decline of Roman metrology after, 9–10
British Standards Institution, 12
Brook, Henry, 41n
Buckingham, 75n
Bury St. Edmunds, 75n
Bushel: of Edgar, the Peaceable, 12, 14; in Treatise on Weights and Measures, 22; in Statute of Purveyors, 23; of brass, 30n; in York, 33; in statute of 1340, 38; urban officials control standards of, 43; standards of, 62, 93; definition of, 77; in troy and tower systems, 78; at London Science Museum, 80; gauging of, 93; Winchester standard of, 95; mentioned, 37, 54. *See also* Standards
Butt, 29, 58

Caesar, Gaius Julius, 3
Calabria, 15
Cambridge, 75n
Cambridge University, 50. *See also* Chancellors
Canterbury, 30
Canute, king of England, 15
Caracalla, Roman emperor, 4
Cardiff, 89
Cardigan, 89
Carew, Richard, 69n
Carlisle, 75n
Carlow, 75n

Carmarthen, 89
Carnarvon, 63, 89
Carnock, 31n. *See also* Sack
Carysfort Committee, 96
Castella, 3
Castle of Dover, 75n
Castra, 3
Catuvellauni, 3
Ceawlin, of Sussex, 10
Celtic tribes. *See* Tribes
Cerdic, of Wessex, 10
Chalder: in statute of 1421, 29; description of, 29; mentioned, 32
Chamberlains: mark of, 47; metrological duties of, 47, 51
Chancellors: metrological duties of, 47, 49–51; at Oxford, 49–50; at Cambridge, 50n, 51n; at Glasgow, 51
Charles II, king of England, 19
Charter Rolls: commissioners appointed in, 41n; metrological duties of monasteries in, 48; metrological duties of bishops and archbishops in, 49
Chaundos, Robert de, 45
Chelmsford, 75n
Cheshire, 47
Chester, 47, 75n
Chester Castle, 47
Chiltenham, William de, 41
Chisholm, H. W.: researches of, 12; on yard standard of 1496, 77n; on Great Fire of 1666, 81n
Chlorus, Constantius, 4
Civil Aviation Authorities, 98
Claudius, Roman emperor, 3
Clergymen, 47
Clericus mercate hospitii regis. *See* Clerks of the market
Clerks, 53
Clerks of the market: in statute of 1389, 26, 68; standards used by, 30n, 68; appointments of, 35; and relations with commissioners, 39, 40, 67; in statute of 1340, 39, 66; in York, 45; loss of power of, 45, 70; restrictions on, 49, 67–69 *passim;* metrological duties of, 59, 60, 65–68 *passim;* for the king's household, 65; sources on, 65–70 *passim;* in Rolls of Parliament, 67; assistants of, 68; fines for infractions set by, 68; height of power of, 68; fines levied on, 69; abuses by, and reaction, 69, 69n; in statute of 1640, 69, 70; repudiation of, 70; avoirdupois weight standards of, 87; troy weight standards of, 87; mentioned, 49, 54, 56
Close Rolls: metrological functions of lords in, 46; metrological functions of monasteries in, 48; port weighers in, 63
Cloths: regulated in Magna Carta, 18; standard dimensions for, in statutes, 23, 27, 59–62 *passim;* repeal and reconfirmation of measurement regulations for, 27–28
Coal, 72
Colchester: site of castrum, 3; commissioners appointed in, 41n; weight standards of, 89
Colford, William de, 41
College of Glasgow, 51. *See also* Chancellors
Comitatus, 9
Commercial Revolution, 71
Commissioners: examination of keels by, 29; abuses by, 43; in competition with clerks of the market, 66; frauds perpetrated by, 67; Exchequer standards supervised by, 75; metrological duties of, 86; examination of standards by, 88; mentioned, 37, 47, 50, 54. *See also* Urban residents
Commissions, 86, 95–96. *See also* Commissioners
Commodus, Roman emperor, 4
Commons, House of: dissatisfaction voiced by, 35; restrictions on clerks of the market by, 67; instructions to clerks of the market by, 68. *See also* Parliament
Composition of Yards and Perches (*Compositio ulnarum et perticarum*): linear and superficial measures of, 20; dating of, 20, 21; significance of, 21; impact of, after thirteenth century, 22–23
Compositio ulnarum et perticarum. *See* Composition of Yards and Perches

Congius, 8
Conservators of the peace, 56. *See also* Justices of the peace
Consumer Council, 98
Convents, 49
Coopers, 82. *See also* Coopers' Guild of London
Coopers' Guild of London: metrological duties and privileges of, 84–85; reasons for loss of powers of, 85n. *See also* Guilds
Cornwall, 38n
Coroners: appointments of, 35; duties of, 52–53; mentioned, 55
Council of Engineering Institutions, 99
Council of the Confederation on British Industry, 79
Courts: staple, 35; tolzey, 35; leet, 35; baron, 35; shire, 35; hundred, 35; circuit, 41; manorial, 46, 47; county, 52; of general eyre, 52; of clerks of the market, 67; of the marshalsea, 68
Coventry: mayor of, 34; standards in, 75n; mentioned, 51
Credit, 71
Crypt Chapel, 15. *See also* Pyx
Cubit, 6, 10
Cubitum, 6
Cumberland, 56
Curia regis, 14
Cusance, William de, 67
Cynric, of Wessex, 10

Daingean, 75n
Dapetot, Robert, 41
Decempeda. *See* Perch
Denarius, 7, 7n
Denbigh, 89
Denmark, 74
Deputy Warden of the Standards, 93
Derby: justices in, 56; commissioners in, 66; standards in, 75n
Devonshire, 54n
Digit, 6
Digitus, 6
Diocletian, Roman emperor, 4
Dirhem, Arabic, 14, 20
Dorchester, 75n
Double actus quadratus, 7

Drapers, 19n
Drogheda, 75n
Dublin: weights and measures standards of, 19, 75n; excess of metrological personnel in, 48; castle of, 65
Dundalk, 75n

East Anglia, 10
East India Company, 74
Eastland Company, 74
Edgar the Peaceable, 11, 12, 14
Edinburgh: standards in, 30; excess of metrological personnel in, 48; statute made in, 51
Edward I, king of England: Composition of Yards and Perches attributed to, 20, 21; Treatise on Weights and Measures of, 20, 21; decree of 1296 of, 22; alnage duties in reign of, 59; yard standard of, 76; mentioned, 65
Edward II, king of England: statute of linear measures of, 22; commissioners appointed by, 38n; urban officials given metrological duties by, 43; ordinance of, 43; death of, 44
Edward III, king of England: statute of 1328 of, 23; statute of 1357 of, 32; iron weights of, 33; administrative procedures during reign of, 38; statute of 1340 of, 39; grant to Ebulo Lesfrange by, 46; bestowal of metrological duties on Oxford University by, 50; protection of oyer and terminer personnel by, 55; alnage legislation by, 59; Close Roll of, 63; clerks of the market appointed by, 66; craft guilds given metrological duties by, 83; weight standards of, 87, 88
Edward the Confessor, 15
Edwin, Bretwalda, 10
Egbert, king of Wessex, 10
Elizabeth I, queen of England: statutes of, 29; standards of, 74–75, 83; Exchequer yard and ell bed, yard, and ell of, 76; withdrawal of privileges of Hanseatic League by, 79; Exchequer avoirdupois weights of, 86; weights

Elizabeth I (*cont.*)
and measures developments under, 86; inception of standards manufacture under, 86; avoirdupois bell-shaped and flat weight standards of, 86, 88; weight standards at Exchequer of, 87; jury impaneled by, 88; Exchequer half-hundredweight standard of, 88; Exchequer troy weights of, 90; Exchequer avoirdupois bronze weights of, 91; yard standard of, 92; capacity measure and Winchester standards of, 93; characteristics of standards of, 93; mentioned, 19n, 81. *See also* Standards

Ell: urban officials control standards of, 43; standards of, 51, 52, 60, 92, 93; mentioned, 37, 61

Ellerker, Nicholas de, 63

Elne, 10, 21

Empire, Roman. *See* Roman Empire

Encyclopédie, 70

England: impact of changing conditions on, in late medieval and Tudor periods, 71–74; new metrological era in, 74; Hanseatic League in, 78–79; distribution of standards in, 89; development of standards after Tudor era in, 95; standards of length in, 96; metric changeover in, 98–99

—Angevin-Plantagenet: commerce in, 16; economic growth of, 17; weights and measures standards and officials of, 17

Ermine Street, 3

Escheator, 68

Essex, 10

Esshewra, Richard de, 41

Esterlingus, 11

Ethelbert, Bretwalda, 10

Ethelbert, king of Mercia, 10

Evesham, 48

Examiners, 82, 97. *See also* Inspectors

Exchequer: beams distributed from, 25n; standards of, 30n, 51, 52, 63, 68, 75, 76, 79, 80–81, 86, 87, 88, 89, 90, 91, 95; Standards Department of, 31n; seals of, 36; failure to deliver fines to, 39; treasurer and barons of, 40, 41; indentures delivered to, 44; metrological duties of officials in, 47, 51; appointments by, 53; receipt of first series of Elizabethan standards, 86; weight standards distributed to, 89; primary reference standards of, 89; usher of, 92. *See also* Standards

Exeter: standards in, 75n, 88; weight standards of Henry VII at, 87; half-hundredweight standard at, 88

Fairs, 55
Fatt, 27, 43
Fees, 66
Feudalism, 9, 46
Fiefs, 14
Fine Rolls, 66
Fines: in Assize of Bread and Ale, 20n; in Ordinance for Bakers, Brewers, and Other Victuallers, 23; in Statute of the Staple, 24; in Statute of Westminster I, 25; in statute of 1405, 28; in statute of 1421, 29; discussion of, 37; in Patent Rolls, 37; in Fine Rolls, 38n; in statute of 1340, 39; for infractions by commissioners, 40; misuse of, 40; in statute of 1328, 44; in mandate to Ireland of 1351, 45; of manorial courts, 46; in *Fleta,* 53; of justices of oyer and terminer, 55; of justices of the peace, 58; collected by alnagers, 60; levied by alnagers, 61, 62; of port measurers and weighers, 63; in statute of 1353, 64; in statute of 1357, 64; collected by clerks and commissioners, 67; limits placed on, 68; inflation of, 69; of clerks of the market, 69; of urban officials, 81; assessed by Coopers' Guild, 84; mentioned, 46

Finger, 6
Firkins: variations in, 82; of ale, beer, and soap, 82; mentioned, 58
Firthindal, 34
Flanders, 16
Fleta, 53
Flint, 89
Foliambe, John, 68

Food Manufacturers' Federation, 98
Foot: of North German system, 10, 11; of Rome, 11; in Composition of Yards and Perches, 21; mentioned, 77. *See also* Pes
Fosse Way, 3
Foster Lane, 83
Fother, 32
Founders' Company of London, 85
France: trade with England, 16; economy of, 71; geographic dimensions of, 72; metric system created in, 96; standards of length in, 96; mentioned, 74
Franks, 5
Frauds: elimination of, 23; in statute of 1340, 23; in Statute of the Staple, 24; in Statute of Westminster I, 25; in statute of 1389, 26; regulation of, 31; among officials, 35; in Fine Rolls, 38n; of turners, 44; of alnagers, 60; investigated by clerks of the market, 65; types of, 75n
Furlong, 11. *See also* Stadium

Gallon: of wine in Treatise on Weights and Measures, 22, 29; in Statute of Purveyors, 23; of wine in statute of 1423, 29; of ale in York, 33; of wine in York, 33; in Beverly, 34; in statute of 1340, 38; definition of, 77; in troy and tower systems, 78; at London Science Museum, 80; standards of, 82, 93, 95; of soap, 83; gauging standard of, 93; mentioned, 37, 54
Gallows, 53
Garwenton, Thomas de, 57–58
Gauger, 26. *See also* Port measurers and weighers
George II, king of England, 95
Germanic tribes. *See* Tribes
Gill, 34
Gird, 12, 13. *See also* Yard
Girda, 13. *See also* Yard
Glasgow, 51
Gloucester: site of castrum, 3; parliament held in, 50; roll of, 55; commissioners in, 66; standards in, 75n
Gloucestershire, 55

Goldsmiths, 88, 89. *See also* Goldsmiths' Company
Goldsmiths' Company: supervision of London's weights by, 83; metrological duties of, 83; weight standards of, 89
Grand juries, 37
Great Famine, 71
Great Fire, of 1666, 81n, 96
Grove Ferry, 13
Guildford, 75n
Guilds: standards of, 63; regulations of, 73; in statute of 1531, 82; metrological duties of, 82–86 *passim*. *See also* Guildsmen
Guildsmen: appointments of, 58; abuses of, 58; metrological duties of, 82–86 *passim*. *See also* Guilds
Gun casting, 73
Gunpowder, 71

Haberjets, 18
Hadrian, Roman emperor, 4
Hadrian's Wall, 4
Halden, Nicholas, 40
Half-bushel: in Statute of Purveyors, 23; in York, 33; standards of, 62; mentioned, 54
Half-hundredweight: definition of, 32n; Exchequer standard of, 88; other standards of, 88, 92
Half-pound, 25
Half-sack, 25
Half-sextarius, 8
Hamburg, 20
Hanseatic League: trade with England, 16; in London, 20, 78–79; monopoly of, 72; merchant's pound of, 78; deterioration of, 79
Hastings, Battle of, 14
Haverfordwest, 89
Hedersete, Simon de, 40
Hemina, 8
Hengest, of Kent, 10
Henle, Walter de, 40
Henry I, king of England, 17
Henry II, king of England: assizes of, 17; entrusted duties to sheriffs, 53
Henry III, king of England: reissue of Magna Carta by, 19; weights and

Henry III (*cont.*)
  measures program of, 19; decrees of, 19; Assize of Bread and Ale of, 19–20, 36, 37; Composition of Yards and Perches attributed to, 20, 21; justices of the peace increase power under, 56; alnage duties during reign of, 59n
Henry IV, king of England: cloth measurement regulations of, 27; motives for 1407 statute, 28
Henry V, king of England, 28
Henry VI, king of England, 29–30
Henry VII, king of England: statutes of, 29; standards of, 74–75, 88; dissemination of standards by, 75, 89; latten metal yard of, 76; Exchequer bronze gallon and bushel of, 79; standards at London Science Museum of, 80; avoirdupois bell-shaped and flat bronze weights of, 80; urban officials in statutes of, 81; weight standards of, at London, Exeter, Worcester, and Norwich, 87; penalties for illegal weights and measures in statutes of, 92; bronze yard of, 92; characteristics of standards of, 93; grain standards of, 93; mentioned, 33, 79
Henry VIII, king of England, 78
Heptarchy, 11
Heredium, 7
Hereford, 49, 75n
Hertford, 75n
Hiltone, John de, 44–45
Hogshead, 58
Hohenstaufen, House of, 71
Holland, 72
Holy Roman Empire, 71
Honorius, Roman emperor, 4
Hopir, 34
Horsa, of Kent, 10
Hundredweight: standards larger than, 32; definition of, 32; mentioned, 45n
Hundred Years War: destruction caused by, 71; impact of, on England and France, 72
Huntingdon, 75n

Ilchester, 75n

Imperial Weights and Measures Act, 96–97
Inch: in Composition of Yards and Perches, 21; standard of, 21n; mentioned, 62, 77. *See also* Pollicus; Uncia
Infractions, 12, 12n
Ingelby, Thomas de, 56
Inquests, 53
Inspectors: metrological duties of, 51, 82, 83, 97; rise to power of, 95; growth in numbers of, 98; mentioned, 59
Ireland: wine measures imported into, 26; urban officials of, 45, 81; clerk of the market in, 65; metrological officers in, 68; cities receiving Exchequer standards in, 75n; English metrological law established in, 76n
Italy, 16

James I, king of England, 85
James III, king of Scotland, 51
John, bishop of Hereford, 49
John, king of England: weights and measures program of, 17; Magna Carta of, 18; mentioned, 34
Joint-stock companies, 74
Jugerum, 7
Juries: empaneling of, 70, 89; duties in connection with standards, 88, 89
Justices: inherit duties from commissioners, 41; metrological duties of, 52; mentioned, 36, 54
—ad hoc, 35
—of assize: metrological duties of, 55; inherit powers of clerks of the market, 70
—of laborers, 55
—of oyer and terminer: in grant of 1355, 55; metrological duties of, 55, 56; inherit powers of clerks of the market, 70
—of the peace: selection of, 35; functions of, 41n, 56, 57; increase in powers of, 56, 58; court sessions of, 57; metrological duties of, 57–59; fifteenth-century statutes on, 58; waning power of, 59; inherit powers of clerks, 70; as overseers, 76; men-

Justices (*cont.*)
   tioned, 46
Justiciar, 65
Justiciary Rolls: metrological duties of monasteries in, 48; metrological duties of archbishops in, 49

Kater, Captain, standards of (Figure 1), 76
Keel: in statute of 1421, 29; chalders in, 29; mentioned, 32
Keel-load, 29
Keeper of weights and measures, 68
Kent: Claudian landing in, 3; establishment of, 10; discovery of weights in, 13; justices of the peace in, 57
Kildare, 75n
Kilderkin: variations in, 82; of ale, beer, and soap, 82; mentioned, 58
Kisch, Bruno, 13
Knight marshal, 65, 65n
Kymberle, Richard de, 66

Lacy, Edmund de, 40
Lacy, Henry de, 46
Lancashire, 73
Lancaster: large measures in, 27; standards in, 75n
Landholdings, 71
Latifundia, 7
League, 6–7
Leaseholds, 72
Lee, John de la, 56
Leicester: commissioners in, 66; standards in, 75n
Lesfrange, Ebulo, 46
Leuga, 6
Levant Company, 74
Lewes, 75n
Libra, 5, 7, 7n, 8
Lincoln: site of castrum, 3; sheriff of, 53, 54; commissioners in, 66; standards in, 75n
Lion of Justice (Henry I, king of England), 17
Lispund, 27
List, of cloth: in Magna Carta, 18; definition of, 18n
London: transfer of Winchester standards to, 15, 17; standards of, 29, 30, 54, 65, 75n, 88; bureaucratic officials at, 35; commissioners appointed in, 37, 39; directive from, 40; appeals to, 40; metrological problems outside urban centers, 42–43; excess of metrological personnel in, 48; court of barons of Exchequer in, 51; sacks of wool shipped to, 55; statutes proclaimed in, 58; master measurers in, 62; weighmaster and weighing machine in, 62–63; increase in population of, 74; standards sent from, 81, 89, 92; Coopers' Guild of, 82, 83–85; metrological duties of guilds in, 83; lead weights of, 85; Pewterers' Company in, 85; avoirdupois weights in, 85; Founders' Company in, 85–86; Corporation of, 86; weight standards of Henry VII at, 87; half-hundredweight standard at, 88; mentioned, 34, 61, 83
Longbow, 71
Lords: metrological appointments of, 35; metrological duties of, 46; inquests at court sites of, 53; mentioned, 47
Lostwithiel, 75n
Lübeck, 20

McCaw, G. T., 12
Magna Carta: of King John, 18; chapter 35 of, 18–19, 19n; capacity measures of, 18, 19; uniqueness of, 19; of 1216 for Ireland, 19; old Saxon elne of, 21; quarter of, 22; weights and measures provisions of, 38n; mentioned, 27
Maidstone, 75n
Manorialism, 46
Marc, 28, 29
Mark: of North German mints, 20; double, of Cologne and Hamburg, 78
Marlborough, 89
Mary, queen of England, 62
Matthew of Paris, 18n
Mayors: presentments before, 35; increase in metrological powers of, 41; metrological duties of, 42–44

Mayors (*cont.*)
  passim; competition by, with university chancellors, 50; barred clerks of the market from cities, 69; coopers appeal to, 84; ordered burning of defective casks, 84; mentioned, 47, 50, 60, 82. *See also* Urban officials
Measurers: of corn, 62; of salt, 62; mentioned 60, 63. *See also* Port measurers and weighers
Measures of capacity: heaped and shallow, 22, 22n; striked, 22n, 23-24
Mediterranean, 73
Mensurae domini regis, 13
Merchant Adventurers company, 74
Merchants, 88, 89
Mercia, 10
Meter, 96
Metrication Board, 99
Metric system: nineteenth-century expansion of, 5; changeover to, 98-99
Middle Ages: Early, 9-10; Late, 71, 83
Mile, 6-7
Mille passus, 6-7
Modius, 8, 34
Monasteries, 48-49
Mongols, 71
Monks, 48-49
Morrow of St. Michael's Feast, 39, 44
Mullingar, 75n
Muscovy Company, 74

Nail, 92
Nessefeld, William de, 56
Newcastle, 75n
Newcastle-upon-Tyne, 29
Nicholson, Edward, 7n
Nomisma, 8
Norfolk, 40
Norman Conquest, 11, 14
Normans, 71
Northampton, 75n
Northumberland, 56
Northumbria, 10
Norwich: standards in, 75n, 88; weight standards of Henry VII at, 87; half-hundredweight standard at, 88
Nottingham: justices in, 56; commissioners in, 66; standards in, 75n

Nuns, 49

Obol, 8
Offa, king of Mercia, 10, 11, 20
Officials, weights and measures: duties of, 34; problems with, 34-35; as part-time personnel, 47; of Tudor era, 81-86; development of, before Stuarts, 94; removal of part-time personnel after Tudor era, 95
O'Keefe, John, 12
Orders-in-council, 74, 86
Ordinance for Bakers, Brewers, and Other Victuallers, 22-23
Ordinance for Measures, 22, 23
Ordinance of 1324, 43
Ostrogoths, 5
Oswald, Bretwalda, 10
Oswiu, Bretwalda, 10
Ottoman Turks, 71
Ounce: of Rome, 7-8; of Offa, 11, 11n; of Arabic silver dirhem standard, 20; avoirdupois, 25, 87; troy, 28, 77, 79, 87, 92; of Swedish mark-weight pund, 28n; of Danish solvpund, 28n; of Scots tron pound, 28n; of Bremen pound, 28n; of Norwegian skaalpund, 28n; of Amsterdam pound, 29n; of Scots trois pound, 29n; of Dutch troy pound, 29n; of French troy pound, 29n; in troy and tower systems, 78; tower, 79
Oxford: weights and measures of, 50; standards in, 75n
Oxfordshire, 54n
Oxford University, 50
Ozingell, 13

Pace, 6, 7
Pack-load, 18. *See also* Quarter
Palm: of Rome, 6; of North German system, 10
Palmipes, 6
Palmus, 6
Pannier, 34
Parliament: growth of, 21-22; lengths of linear measures legalized by, 22; pressure of Lancastrians on, 27; contradictory cloth measurement regulations of, 27-28; standardiza-

244 Index

Parliament (*cont.*)
tion aims of, 35; shire commissioners established by, in 1340, 38–39; commissions repealed by, 39–40; metrological programs of, for urban centers, 42–43; cloth regulations established by, 44, 60; mandate sent to Ireland from, 45; clerks of the market restricted by, 45, 68, 69, 69n, 70; bestowed metrological duties for various reasons, 48; rolls of, 55; duties assigned to justices by, 57; port measurers and weighers restricted by, 63–64; control over taxes by, 72; construction of new standards ordered by, in reign of Henry VII, 75–76; tower pound abolished by, 78; construction of capacity measures restricted by, and resulting problems, 82–83; new yard standard authorized by, 93; summary of metrological programs of, after Tudor era, 95–97; mentioned, 52, 59, 62n, 88

Passus, 6. *See also* Pace

Patent Rolls: appointments of metrological personnel in, 37; urban residents' duties in, 38; appointments of commissioners in, 40; problems of appointments in, 40–41; metrological duties of urban officials in, 43; university chancellor of Oxford in, 49–50; appointments of justices in, 56; port measurers and weighers in, 63; clerks of the market in, 66

Pax Romana, 4

Peck: in Statute of Purveyors, 23; in York, 33; mentioned, 54

Pendulum, 95, 96

Penny, 11, 11n. *See also* Pennyweight

Pennyweight: of Offa, 11, 11n; of troy pound, 28; definition of, 77; in troy and tower systems, 78. *See also* Denarius

Perch, 6, 21

Perth, 30

Pertica, 6. *See also* Perch

Pes, 5, 6. *See also* Foot

Pewterers' Company of London: infractions by members of, 44–45; given metrological duties, 85

Pfund, 20. *See also* Pound

Philippa, queen of England, 56

Pike, 71

Pilkyngton, John de, 68

Pillory, 53

Pint, 34

Pipe: in fourteenth-century statutes, 26; in statute of 1423, 30; mentioned, 58

Pius, Antoninus, 4

Plautius, Aulus, 3

Plumbers' Company of London: given metrological duties, 85; fees of, for assays, 85n

Poland, 74

Pollicus, 6. *See also* Inch

Pondi regis, 13

Port Laois, 75n

Port measurers and weighers: metrological duties of, 26, 59, 62–65; sources on, 62, 63, 64–65; standards of, 62, 63, 64; injunctions against, 63; infractions by, 64; rules and oath of, in Ordinance Book of 1565, 64

Postlethwayt, Malachy, 15n

Pottle: in Statute of Purveyors, 23; of ale and of wine in York, 33; in Beverly, 34; mentioned, 54

Pound: of Rome, 7–8; in Statute of Westminster I, 25; in troy and tower systems, 78; mentioned, 46. *See also* Libra

—avoirdupois: before Elizabeth I, 25; commercial pounds replaced by, 79; standards of, 87, 92; mentioned, 83

—Bremen, 29

—commercial: of southern Germany and Scotland, 78; of 7680 and of 7200 grains, 78; eradication of, 79

—Dutch, 29

—mercantile, 78, 83

—merchant's, of Hanseatic League, 78–79

—moneyer's, 11

—Norse, 27

—North German, 28

—Old Etruscan, 7

—Oscan, 7

Pound (*cont.*)
—Saxon, 11, 11n, 13
—Scots, 29
—ship, 27
—tower: in Assize of Bread and Ale, 20, 21; in Treatise on Weights and Measures, 22, 77; in Ordinance for Measures, 23; compared with troy pound, 28; used as weight content of wine gallon, 29; abolition of, 78; uses of, 78; mentioned, 11n
—troy: earliest mention and use of, 28; definition of, 28, 77; derivation of name, 28–29; replaces tower pound, 78; standards of, 96; abolished, 97; mentioned, 83, 87
Poutrel, Richard, 56
Priors, 35. *See also* Monasteries
Pyx: Chamber of the, 15; Chapel, 15, 15n; trials of the, 15, 83

Quart: in Statute of Purveyors, 23; of ale and of wine in York, 33; in Beverly, 34; mentioned, 54
Quartarus, 8
Quarter: in Magna Carta, 18, 19; definition of, 19, 54; standardized in 1296, 22; in cloth trade, 23, 60; heaped, 27; of London, 43; standards of, 62
Quarter-hundredweight, 32n
Quarter-pound, 25
Quarter-sack, 25
Queen's Hospital, 89

Ravenser, Richard de, 56
Roy, General (Figure 1), 76
Reading, 75n
Reeves, 38
Richard I, king of England: assizes of, 17; Assize of Measures of, 18; weights and measures decree of, 18; urban residents receive metrological duties under, 35; office of coroners created by, 52; justices of the peace under, 56
Richard II, king of England: weights and measures legislation during reign of, 26–27; last statutes of reign of, 27; metrological duties bestowed on Cambridge University by, 50; charter to craft guilds of, 83; mentioned, 33, 68
Richemond, Peter de, 56
Rochester, 45
Rod: of North German system, 10, 11; in Composition of Yards and Perches, 21. *See also* Perch
Roman Empire: Britain invaded by legions of, 3; Germanic invasions of, 5; commerce, industry, and agriculture of, 8; population of, 8
Rotl, Arabic silver, 11
Roy, General, standard of (Figure 1), 76
Royal Society, 76, 99
Rural aristocracy: metrological duties and functions of, 46–47; metrological records on, 46–47. *See also* Lords
Russets, 18

Sack: in Statute of Westminster I, 25; in statute of 1389, 27; equivalent to carnock, 31n; of wool, 32, 55; definition of, 55
St. Anthony's Cross, 84
St. Botolph, 31n
St. Michael's Feast, Morrow of, 39, 44
St. Thomas of Hereford, 49
Salisbury, 75n
Salters' Company, 62
Scales: maintenance of, 64; used by Plumbers' Company of London, 85. *See also* Balance; Beam
Scandinavia, 74
Schoenus, 7
Science Museum, London: standard yard of 1497 preserved at, 76–77; capacity measure standards at, 80; description of avoirdupois standards at, 80–81; Exchequer avoirdupois weights deposited at, 86, 89, 91; Exchequer troy weights deposited at, 90; standards of 1588 at, 92. *See also* Standards
Scotland: highlands of, 3; trade with England by, 16; weights and measures of, 51; metrological duties of urban officials in, 81; metrological

246   Index

Scotland (cont.)
  duties of coopers in, 85
Scruple, 8
Scrupulum, 8
Sealers, 62. See also Searchers
Seals: of William I, 13, 15; in Ordinance for Bakers, Brewers, and Other Victuallers, 22, 23; forms and applications of, 33n; of Exchequer, 36; in ordinance of 1324, 43; of alnagers, 60–62 passim; Tudor standards stamped with, 76; of coopers, 82
Seam, 18. See also Quarter
Searchers, 60, 62
Sears, J. E., 93
Sergeants, 53
Severus, Septimius, 4
Sextarius, 5, 8
Sextula, 8
Sheep raising, 72
Sheffield, 73
Sheriffs: orders to, following Statute of Purveyors, 24n; balances and wool weights sent to, 32; metrological duties of, 51, 52–54; characteristics of, 54; police work of, 56; warrants to, 70
Shippers' Council, 98
Shirburn, Thomas de, 67
Shrewsbury, 75n
Sicily, 15
Signs of coopers, 84
Siliqua, 8
Skinner, F. G., 13
Skippund, 27. See also Pound
Somerset, 73
Southampton: sheriff of, 54; standards in, 75n
Spynes, Richard, 40
Square actus, 7
Squires, 73
Stade, 6
Stadium, 6
Stafford, 75n
Standards: construction of, 30; regulation of, 30–31; distribution of, 30n, 31n; problems with, 31; copies of, 31; deterioration of, 31, 69; surviving from pre-Tudor times, 32; of 91-pound wool weight, 32; in Assize of Bread and Ale, 36; of Scotland, 51; of master measurers, 62; of king, 65, 66; of Tudors, 74–81; of English linear measures (Figure 1), 76; in statute of 1496, 77; of Exchequer bronze gallon and bushel (Figure 2), 79; of avoirdupois weights (Figure 3), 80; of Exchequer avoirdupois weights (Figure 4), 86; of Elizabeth I, 86–93; of weights deposited at various locations, 87; of Exchequer avoirdupois bell-shaped gunmetal weights (Figure 5), 89; of Exchequer troy bronze weights (Figure 6), 90; of Exchequer avoirdupois bronze weights (Figure 7), 91; proclamation of, 92; of length, 92–93; of capacity measures, 93; development of, before Stuarts, 94; rapid success in manufacture of, 95; of troy pound, 96; of yard, 96; development between 1790 and 1824, 96; impact of twentieth century on, 97–98
Standards Commission, 96
Standards Department: of Exchequer, 31n; of Board of Trade, 80
Standing Joint Committee on Metrication, 99
Standyssh, Ralph de, 68
Statute of Purveyors: contents of, 23; privileges granted in, 24; balance mentioned in, 24n, 25n
Statute of the Staple, 24
Statute of Westminster I, 24–25
Statutes of laborers, 57
Steelyard, 32, 79
Sterling, 11, 77
Sterlingus, 11
Stewards: presentments before, 35; of manorial courts, 46
Stone: in statute of 1389, 27; of wool, 32; definition of, 55; mentioned, 46
Stowe, Humphrey de, 66
Streeter Collection, 13
Strike: of Coventry, 34; standards of, 62
Stuarts, 94
Sub-beadles, 53

Sussex, 10
Switzerland, 71
Sworn inquests, 37

Talent, of Alexandria, 7n
Tenants-in-chief, 14
Tertian, 30, 58
Textiles, 73. *See also* Cloths
Theodosius, Roman emperor, 4
Thumb, 10
Ton, 32
Tower of London: standards constructed in, 30n; Wardrobe of, 31n; yard standard of, 76; avoirdupois and troy weight standards in, 87, 87n, 89
Tractatus de ponderibus et mensuris. See Treatise on Weights and Measures
Treasury, 75
Treatise on Weights and Measures *(Tractatus de ponderibus et mensuris)*: dating of, 20; contents of, 22; influence on Ordinance for Measures, 23; problems of wording in, 77; tower pound in, 77
Tribes: Celtic, 3, 4; Germanic, 3, 4, 10
Trim, 75n
Trinovantes, 3
Trona, 25n
Tronator, 25n, 33
Trone, 25n. *See also* Beam; Balance
Troner, 33. *See also* Tronator; Tronour
Tronour, 25n. *See also* Tronator, Troner
Tudors: succession of, 72; impact of mercantile classes on, 72; rapid economic gains under, 72-73; major metrological contributions of, 74; standards of, 74, 75; metrological growth patterns after, 94
Tumbrel, 53
Tun: in fourteenth-century statutes, 26; in statute of 1423, 30; mentioned, 58
Turk, Walter, 44-45
Turners: standards constructed by, 31; of London, 44; duties of, 44n

Ulna: in Magna Carta, 8; definition of, 18; equivalent of Saxon elne, 19; in Composition of Yards and Perches, 21
Uncia (inch), 6
Uncia (ounce), 7, 7n, 8
Under-sheriffs, 53
Uppingham, 75n
Urban officials: metrological duties and growth of power of, 42-45, 81; in ordinance of 1324, 43; interaction with commissioners, 43; in statute of 1328, 44; of Tudor era, 45; resentment of, by Chancellor of Oxford University, 50; height of powers of, 81; in statutes of 1491 and 1495, 81; in Ireland and Scotland, 81; decline of metrological powers of, 81-82; mentioned, 83. *See also* Aldermen; Baliffs; Mayors
Urban residents: metrological duties of, 35-42; in Assize of Measures, 36; in Assize of Bread and Ale, 36-37; in Patent Rolls, 37-38; in statute of 1340, 38-39; in Fine Rolls, 38n; problems with, after statute of 1340, 39; metrological duties repealed, 40; appointments of, 41; reappointed by statute of 1351, 41; decline of metrological duties after 1351, 41-42. *See also* Commissioners
Urna, 8
Vandals, 5
Vassals, 46. *See also* Rural Aristocracy
Verge (measure), 61. *See also* Yard
Verge (territory): definition of, 67; of clerks of the market, 68; of the king's court, 69
Viscounts, 55
Visigoths, 4, 5

Wales: wine measures imported into, 26; clerk of the market in, 66; distribution of standards in, 89; mentioned, 65
Wardens: of Coopers' Guild of London, 84; of Pewterers' Company of London, 85; of the Standards, 88n, 92
Wardrobe: of Tower of London, 31n; fines brought to, 67
Warener, William, 56
Wars of the Roses, 72

Warwick, 66
Waterford, 45
Watling Street, 3
Weighers, 63. *See also* Port measurers and weighers
Weights: in statute of 1340, 38; assay of, 46; of lead, 85; standards of, 87; mentioned, 37. *See also* Ounce; Pound; Standards
—bell-shaped, 13. *See also* Henry VII, king of England; Elizabeth I, queen of England
—disk, 13
—of tron system, 51
—avoirdupois: uses of, 78; Founders' Company's control over, 85; used by clerk of the market for the king's household, 87; standards in the Tower of London of, 87; standards of, 87, 89, 92
—mercantile, 78
—tower: comparison with troy, 78; supervision of, 83
—troy: comparison with tower, 78; effects of statute of 1496 on, 78; supervision of, 83; used by clerk of the market for the king's household, 87; standards of, in Tower of London, 87; standards of, 89, 92, 93
Weights and measures: growth patterns and units of, before Stuarts, 94; aims of reformers of, after Tudor era, 94–95; reform plans and programs for, after Tudors, 95; problems of unit standardization after Tudor era, 95; effects of scientific experiments on, 96; major characteristics after 1824, 97; major problems of, in nineteenth century, 97; impact of twentieth century on, 97
Weights and Measures Act: of 1878, 97; of 1963, 98
Wessex, 10
Westgate Museum: standards in, 32; original leaden seals of, 33
Westminster: council at, 57; standards in, 75n

Westminster Abbey, 15
Westmoreland, 56
Wexford, 75n
Wey, 32
Whetlay, John de, 41
William I, king of England: governmental changes made by, 14–15; weights and measures policy of, 15; weights and measures decree of, 15, 15n, 16; Winchester standards of, 17; seals of, 23. *See also* Britain, Norman
Wiltshire, 54
Winchester: measures of, 11–14 *passim;* standards of, 12, 15–16, 75n, 88; avoirdupois weights of, 32; Great Fair at, 32–33; bureaucratic officials at, 35; weight standards of Edward III at, 87; half-hundredweight standard at, 88; coal bushel of, 95. *See also* Standards
Winchester Castle, 54
Witan, 9, 14
Wivyll, Henry de, 66
Wool: exports of, 72; increase in price of, 73
Worcester: standards in, 75n, 88; weight standards of Henry VII at, 87; half-hundredweight standard at, 88
Writs, 53
Wychingham, Geoffrey de, 44
Wyldgrys, J., 51–52

Yard: of Saxon kings, 12; at Winchester, 13, 13n; in Composition of Yards and Perches, 21; in statute of 1328, 23; standards of, 51, 76–77, 92, 93, 96; Exchequer standard of, 51, 52; of alnagers, 61; imperial standard of, 92; mentioned, 62
Yeomen, 73
York: site of castrum, 3; standards of, 33, 75n; urban officials in, 45; justices in, 56; commissioners in, 66
Yorkshire, 54

COMPOSED BY FOX VALLEY TYPESETTING, MENASHA, WISCONSIN
MANUFACTURED BY MCNAUGHTON & GUNN, ANN ARBOR, MICHIGAN
TEXT AND DISPLAY ARE SET IN TIMES ROMAN

Library of Congress Cataloging in Publication Data
Zupko, Ronald Edward.
British weights and measures.
Bibliography: p.
Includes index.
1. Weights and measures—Great Britain—History.
2. Mensuration—Great Britain—History. I. Title.
QC89.G8Z86   389'.0942   77-77430
ISBN 0-299-07340-8